Management for Professionals

More information about this series at http://www.springer.com/series/10101

Leigh-ann Onnis

HRM and Remote Health Workforce Sustainability

The Influence of Localised Management Practices

 Springer

Leigh-ann Onnis
College of Business, Law & Governance
James Cook University
Cairns, QLD, Australia

ISSN 2192-8096   ISSN 2192-810X   (electronic)
Management for Professionals
ISBN 978-981-13-2058-3   ISBN 978-981-13-2059-0   (eBook)
https://doi.org/10.1007/978-981-13-2059-0

Library of Congress Control Number: 2018952622

This Springer imprint is published by the registered company Springer Nature Singapore Pte Ltd.
The registered company address is: 152 Beach Road, #21-01/04 Gateway East, Singapore 189721, Singapore

Foreword

To be a remote area nurse, or in fact any remote or rural health professional, you need to have a sense of adventure, think of yourself as resistant and be confident that you can turn your hand to just about anything. Some may think that is pretty haughty; however, it is exactly these traits that procures health professionals into some of the toughest and loneliest parts of Australia (and the world) to ensure access to healthcare for some of the neediest within our society.

I worked as a rural nurse, remote area nurse and have also undertaken missions with the international Red Cross in Wars and disasters across the globe. After 10 years of this exciting and altruistic lifestyle, I wanted to settle back in regional Australia. I didn't want to do shift work so management seemed like the easiest option, it was just management, how hard could it be!

Learning by trial and error, I was fortunate to slowly gain some of the skills to manage a successful team and service within aged care, small rural hospitals, remote health centres and eventually at a regional and national level. The skills, knowledge, intuition and personal behaviours far exceeded what I expected and made the memory of providing remote clinical care seem stress-free!

Many of my contemporaries have very similar histories. Being the last one left standing often meant that you become the manager by default, with very little consideration of ability or suitability, placing the manager, their staff and the health service at increased risk. CRANAplus as the professional body for remote health has for many years identified this as a key challenge within our industry.

Working in such a complex area, with isolation, skill mix, community demands, poorer health outcomes, workforce shortage all while under the constant gaze of policymakers, community and our urban colleagues means that we need the best, most skilled managers, not the least prepared. It is no wonder that historically we have thrown so many great remote health professionals under the 'proverbial bus' when we force management responsibilities upon them.

HRM and Remote Health Workforce Sustainability—The Influence of Localised Management Practices is a long overdue resource for the remote manager work-force that clearly, concisely and practically helps managers address the challenges and celebrate the successes associated with providing sound human resource management to the remote and isolated workforce.

Based on meticulous research and the participation of more than 200 remote health professionals, this book is easy to read and fashioned to the unique context of remote health within Australia. This is an essential resource created specifically for the Remote Health Manager, written by someone who has lived and breathed the realities of Remote HRM and continues to this day to be an expert in our area of practice.

I have had the pleasure to know and work with Leigh-ann Onnis in a variety of professional capacities over the years. Her knowledge is deep, her practice sound, her solutions innovative, her research applied and her mannerisms reassuring. On behalf of the remote health Industry, I commend Leigh-ann for creating such a useful and enduring resource and highly recommend all remote health managers read and keep a copy of this book handy.

Melbourne, Australia   Christopher Cliffe
                                                              Remote Area Nurse
                                           Chief Executive Officer, CRANAplus
                                              Chair of the Coalition of National
                                            Nursing & Midwifery Organisations

Preface

It was a cold Melbourne morning in August 2003 when we set off on our adventure around Australia. An adventure that did not just take us to faraway places; those places shaped our life in ways that we never imagined. As the dust filtered through our lungs and the pindan pumped through our veins, we felt a connection to remote Australia that kept us from returning to our city life. Now living in regional Australia, this book emerges from a career in Human Resource Management and my experiences working with remote clinical teams. During my time working with health professionals in regional and remote Australia, I saw well-intention, but ill-prepared, health professionals eagerly taking on immense challenges. They showed resilience and strength in difficult situations and often accepted poor physical working conditions. For many years, I sat by the 'company turnstile' welcoming the parade of optimistic, enthusiastic health professionals as they headed off for their remote adventure; and supported, then debriefed overwhelmed, exhausted and dispirited health professionals as they left the organisation. During this time, I wondered why organisations experiencing high turnover didn't do more to support the managers and health professionals who are already in remote regions.

This curiosity was the driver for the research studies, and for the past few years it has been a pleasure to meet many of the dedicated, hard-working and remarkable people who provide health services to people in remote regions, often in the most difficult of circumstances. It was a privilege to hear their stories and the research study led to a Ph.D. thesis and series of academic publications and presentations reporting the findings. However, the passion behind this research was always to help managers who work in remote regions, managers who are doing their best to manage remote health workforces with varying levels of experience, management training and support. Therefore, this book is written for all the managers, and aspiring managers, learning to managing people amidst the daily challenges and rewards of providing health services in geographically remote and isolated regions.

Cairns, Australia   Leigh-ann Onnis

Acknowledgements

To my husband Rob, thank you for your unwavering support.

To my friend, and former remote clinician Linsey, thank you for all the conversations that mattered.

To the remote health professionals, health managers and HR managers who participated in the research study, as well as those who shared stories with me at conferences and workshops; thank you for your overwhelming support. This book tells your stories and I thank you for the privilege of being able to communicate your collective voice.

Contents

Part I Remote Health Workforce Sustainability

1 **About the Book**   3
  Introduction   3
  Remote Health Managers   5
  Remote Health Workforces   7
    Indigenous Peoples   8
  Defining 'Remote'   8
    The Distinction Between Rural and Remote   9
    International Remote Contexts 10
  How to Use to This Book 11
  Why Does Workforce Sustainability Matter? 12
  What Does the Human Resource Management (HRM) Approach
  Offer? 15
  References 17

2 **Remote Workplaces** 21
  Synthesis of the Known Rewards and Challenges 22
    Person-Fit (Personal) 23
    Health and Wellbeing (Personal) 23
    Differences Between Professions (Professional) 24
    Professional Development (Professional) 24
    HRM Policy Choices (Organisational) 25
    Management Practices (Organisational) 25
    Environment, Climate and Culture (Contextual) 25
    Rewards 26
  What Can We Learn from the Synthesis of the Literature? 27
  HRM and the Remote Workplace 30
  Decision Making in the Remote Workplace 32
  Managers Are the Key to Workforce Sustainability 33
  References 34

3 **Sustainable Remote Health Workforces**   37
   What Is Workforce Sustainability?   37
   Perspectives from the Current Remote Workforce   39
      Research Findings About the 'People' Aspects of Workforce
      Sustainability   40
      Research Findings About 'Practice' Aspects of Workforce
      Sustainability   45
      Research Findings About 'Place' Aspects of Workforce
      Sustainability   48
   Towards Sustainable Remote Health Workforces   50
   References   51

Part II Human Resource Management Challenges

4 **Recruitment: Attraction, Advertising and Realistic**
   Recruitment   55
   When It Feels like Recruitment Is All You Do   55
   Recruitment Advertising   56
   Psychological Contracts   59
   Communicating with Potential Employees   60
   Recruiting Managers   62
   Transition from Clinician to Manager   63
   Attraction   67
   Realistic Recruitment   68
   Local Recruitment   70
   Expectations   72
   Effective Recruitment Practices   73
   References   74

5 **Remuneration: Extrinsic and Intrinsic Rewards, Incentives**
   and Motivation   77
   Remuneration as Compensation   77
   Financial Incentives   78
      Incentives   79
      Retention Incentives   80
      Policy Conflicts   82
   Perceived Inequities   83
   Motivation and Job Satisfaction   86
   Workforce Sustainability Is Contingent on Both Intrinsic
   and Extrinsic Rewards   89
      Herzberg's Motivation-Hygiene Theory   90
   References   91

**6   Relationships: Social Exchanges, Community Ties
     and Employee-Manager Relationships**   93
     Professional and Personal Isolation   93
     Peer Support   95
     Contemporary Employment Relationships   97
     Employee-Manager Relationship   99
        The Manager-Employee Relationship Influences Organisational
        Commitment..   101
        A Shared Understanding   102
     References   105

7   Resourcing: Access, Availability, and Localisation   107
     Implications of Poor Resourcing   107
     Accommodation   109
     Workplace Health, Safety and Wellbeing   111
     Localised Management Practices Shape Workforce Sustainability   115
     References   116

Part III   Localising Management Practices

8   An Integrated HRM Framework for Remote Managers   121
     Reframing and Refocusing   121
     Tailoring Training...   122
     Supportive Management Practices   123
     A HRM Framework for Remote Managers......................   124
     An Integrated HRM Framework for Sustainable Remote Health
     Workforces   129
     How Can the *I*-HRM-SRHW Assist Remote Managers?   131
     References   132

9   Practice Is Everything...................................   135
     Why Practice Is Everything...................................   135
     Context Is Everything.......................................   136
     The Contextual Challenges of Putting HRM Policies into Practice.....   136
     Managers Are the Key to Workforce Sustainability: The Next
     Generation   139
     The Remote Manager's Journey Continues....................   140
     References   140

Appendix A   143

Appendix B   145

Appendix C   151

Appendix D   155

Appendix E 157

Glossary 161

Further Reading and Resources for Remote Managers 163

Part I

Remote Health Workforce Sustainability

About the Book

Success in management requires learning as fast as the world is changing.

Warren Bennis, Author and Scholar

Key Messages

- This book is written for health managers and aspiring health managers working in geographically remote regions across the globe.
- Managers need to consider the context; because context is important.
- Managers need to localise their management practices by having the skills, capacity and the confidence to draw on their own knowledge and experience to find their own situation-specific solutions.
- Human Resource Management (HRM) is the development of everything within an organisation that supports effective management practices.
- Managers are the key to workforce sustainability in geographically remote regions.

Introduction

'Inspiring leaders know themselves and their capabilities. They dedicate time each day to reflecting on their decisions and evaluating their performance critically and honestly.'

(Robertson 2014, p. 276)

Management is described as the purposeful organisation and coordination of activities in order to achieve an organisation's objectives and vision. Management is usually distinguished from leadership, with Management Consultant Peter Drucker credited with saying that *management is about doing things right, while leadership is about doing the right thing*. Therefore, management is most frequently

© Springer Nature Singapore Pte Ltd. 2019
L. Onnis, *HRM and Remote Health Workforce Sustainability*,
Management for Professionals, https://doi.org/10.1007/978-981-13-2059-0_1

concerned with strategy, culture, objectives, governance and accountability. However, for many new managers it is difficult to find the information that tells them what to do in the specific management situation they face. Over time, managers come to realise that there is no 'checklist' when it comes to managing people, no guide to get through any situation because as managers that is your role. Managers need to determine the best course of action in the given situation, considering the competence, skills and experience of their team, and the available resources. Furthermore, managers need to consider the context; because context is important. For managers in geographically isolated and remote locations, context is everything!

Using Human Resource Management (HRM) theories, this book focuses on the influence of localised management practices on health workforce sustainability. This book is written for health managers and aspiring health managers working in geographically remote regions across the globe. The book contains ideas and information gathered through three related research studies, and as such, an academic evidence-base supports the foundation for this book. However, with the remote manager in mind, every effort has been made to reduce the amount of academic and management jargon. Instead, an overview of the research methodology and HRM theories are provided in the appendices which provide reference points for the information contained in the chapters.

The content of this book is an extension of three separate, yet related, research studies and provides information tailored to the needs of remote managers. The aim is to provide health managers with an evidence-based resource that can guide their management practice. It does not aim to be another academic publication, with a well-known story, along a well-trodden path. In writing for the remote manager (or aspiring remote manager), the book offers wisdom from health managers, Human Resource (HR) managers and health professionals working in remote regions to complement the academic information offered. More importantly, each chapter includes guidance around people management challenges, including practice examples and suggestions about resources to create your own 'Manager's Toolkit'. The book argues that managers need to localise their management practices. This means that managers must have the skills, capacity and the confidence to draw on their own knowledge and experience to find their own situation-specific solutions. These solutions must be consistent with contemporary practice, within the acceptable boundaries of their profession, and compliant with organisational governance; yet also suitable and appropriate for their remote workplace. For managers to be able to adapt to situations and know how to respond with a localised solution, they need to be confident with their own management and leadership style. Hence, throughout this book managers are encouraged to create their own Manager's Toolkit of resources to draw on depending on the situation they face. The toolkit is personalised and can guide each manager's development of their own localised approach to management. How you as a manager develop your own personalised toolkit will become clearer as your progress through the book.

Therefore, this book will be of interest to managers and aspiring managers, working or planning to work in geographically remote regions worldwide, in two ways:

– it provides insight into the workforce challenges and rewards of remote management practice, reinforcing that you are not alone with the challenges you are experiencing; and
– it provides resources and management tools that you can use to create your own Manager's Toolkit.

For Human Resources Professionals, and senior managers, this book is a resource that can be used to build management capacity, enhance current management skills, and as part of management development programs. Particularly, those tailored to the needs of remote managers.

The book is divided into three sections. The first section introduces the book and examines workforce sustainability. The second section focuses on the HRM challenges and looks more closely at the four themes that emerged through the research: recruitment, remuneration, relationships, and resourcing. The third section, localising management practices, introduces the HRM Framework developed for remote managers and then uses the HRM framework to suggest ways in which remote managers can direct their efforts towards improving workforce sustainability. The book concludes by pulling together the theory, stories, the HRM framework and the activities spread throughout this book; then suggesting a pragmatic way forward for remote managers.

Remote Health Managers

'The iconic boab trees of the Kimberley, Western Australia, symbolise strength, resilience and the interdependence of various ecologies; characteristics synonymous with leadership. Extremely resilient to harsh and changing environmental conditions… [The boab] can adapt and thrive by living in mutually beneficial ways with its social and environmental ecologies.'

(Zuber-Skerritt et al. 2015)

In many ways, health managers based in geographically remote regions (hereafter called remote managers) can be compared with the boab, for it is their resilience and their strength that often sees them thrive in the harshness of remote Australia. The boab (Fig. 1.1) provides a metaphorical image of the remote manager. Even in the most precarious of geographical environments, the boab tree can thrive, and so too can the remote manager. Boabs 'grow deep roots for many years, only spreading their limbs above ground once they are firmly established' (Zuber-Skerritt et al. 2015, p. 90). Similarly, remote managers must establish themselves as managers and have a solid foundation before they can grow and perform to the best of their abilities in geographically remote regions. Nature is best in balance; so too are remote managers.

Fig. 1.1 The Boab (Baobab)
(*photo credit* Leigh-ann
Onnis)

Often, finding this balance is challenging for remote managers, especially considering the challenging environment in which they work. Despite these challenges there are many remote managers who thrive and others that are on their way to being the managers that they aspire to be in their remote community. This book focuses on the remote manager, the frontline manager who sits between the health workforce and senior management. Why? Well, the central argument of this book is that these frontline managers are the key to any sustainable change effort. If remote health services want to improve the quality of health services in remote regions worldwide, the frontline manager is the key to successful implementation and any sustained effort to maintain service improvements.

One of the many factors that contribute to quality health services is a competent and accessible health workforce. However, finding sustainable workforce solutions in geographically remote regions has proven to be a difficult task for some health service organisations. Yet, others notice that there are some areas where the workforce is quite stable and more often than not, that stability is attributed to aspects of the manager's leadership style and localised management practices. Therefore, ensuring that competent and experienced managers are supported to both personally and professionally develop in geographically remote regions is essential. But, this is not as easy as it sounds for many reasons. This book will explore these reasons in more detail building on the findings from three research studies that examined aspects of workforce sustainability in tropical northern Australia (hereafter called 'the study').

Remote Health Workforces

The remote health workforce is a subset of the entire health workforce and faces the same issues of global workforce shortages with the added challenges of geographical remoteness (Ono et al. 2014; WHO 2010). Some of these challenges arise from the gender imbalance experienced in caring professions; the hierarchical nature of the professionals that comprise the healthcare systems globally; the disparity between the desirability of urban, rural and remote localities; and the increased mobility of health workers internationally (Hawthorne 2001; HWA 2014; Lippel et al. 2017). An example of how these comparable global issues compound the challenges can be seen in many countries. In Australia, 89.7% of nurses and 98.2% of midwives are female (Hawthorne 2001; HWA 2014). Furthermore, within this national nursing workforce, approximately 82% of the Australian remote health workforce is female (HWA 2014). With more internationally trained female nurses and midwives migrating to Australia than internationally trained female doctors the gender balance between professions will continue to widen (Hawthorne 2001).

Beyond gender, the increased proportion of overseas trained GPs working in remote areas (47%) will continue to grow as government policies require migrating doctors to undertake rural/remote placements before visas are issue to work in cities (HWA 2012). Coupled with a high proportion of international trained nurses working in remote regions, there are obvious challenges with mobility in remote regions. In addition, the maldistribution of health professionals further exacerbates the challenges associated with mobility (Newhook et al. 2011). In Canada, 11.8% of their regulated nursing workforce provides care for their rural and remote population (MacLeod et al. 2017). Overall, the geographic maldistribution of nurses in Canada is reflective of the distribution of health professionals worldwide (MacLeod et al. 2017). An OECD report on the global situation reported that the 2012–2013 OECD Health System Characteristics Survey revealed that only one country (Netherlands) out of 30 OECD countries considered the distribution of doctors between urban, rural and remote regions not to be an issue (Ono et al. 2014).

In Australia, the remote health workforce forms approximately 2% of the Australian health workforce (HWA 2014). Although the remote workforce may comprise only a small proportion of the health workforce, these health professionals work across a large geographical area, and usually work to the full scope of their clinical practice often with limited resources (Bent 1999; Hegney et al. 2002a, b, c). As Health Workforce Australia (HWA) (2012, p. 21) explained:

'delivery in rural and remote settings can vary significantly to that in urban areas. For example, in a rural and remote area a doctor is more likely to deliver health services across acute, aged care and community settings and across traditionally separate professional disciplines, whereas in an urban setting, people often visit specialists within each setting and/or discipline.'

Therefore, the workforce in remote areas must be flexible and highly skilled as they are often required to work in multi-disciplinary teams within the full scope of their professional practice. This is quite different to the way that most urban health professionals practice and can lead to further complexities with the remote workforce, particularly around role clarity and role delineation (HWA 2012; Onnis and Pryce 2016). Consequently, managing health workforces in remote areas necessitates management practices congruent with the working conditions. This includes considering the compounding challenges of gender imbalance, the hierarchical nature of healthcare systems, the lower desirability of many remote localities and the increased mobility of health workers internationally. Health managers may be located in remote regions where they are professionally and geographically isolated; or they may be managing by distance, at times from regional or metropolitan cities hundreds of kilometres away with limited understanding of the context in which remote health professionals work. Both scenarios require management practices that ensure that the workforce is managed effectively so that health services remain accessible for remote populations.

Indigenous Peoples

The author recognises the diverse histories of the world's Indigenous peoples, particularly the history of Australia's Aboriginal and Torres Strait Islander peoples and respectfully refers to them in this book as Indigenous peoples (Australian Human Rights Commission 2005; Torres Strait Islander Regional Council 2016). This does not in any way seek to diminish their respective histories and does not assume in any way that they are one peoples. The study that underpins this book did not collect data that specifically identified Indigenous people or provide any specific findings that related solely to Indigenous Australians. However, there were Indigenous participants, and as such, their voice is contained within the professional perspectives offered throughout this book.

Defining 'Remote'

There are several ways to determine remoteness. As 'remote managers' are the focus of this book it is important to define what 'remote' means in the context of the book. This is being raised for four reasons: (1) to be transparent about the method used in the underpinning studies; (2) to acknowledge the different ways in which 'remote' is defined in Australia; (3) to acknowledge that for health professionals the distinction between rural and remote may be different to that of researchers and statisticians; and (4) to examine international definitions of remoteness to improve understanding about the extent to which the findings from Australia are comparable with similar geographical conditions internationally.

Remote and rural regions universally have fewer residents dispersed across a wider geographic area than cities where the population is denser. In their analysis of the OECD regional database, Ono et al. (2014) found that in Iceland, Australia, Canada and Norway, the population in rural regions is less than 10 people per kilometre; while Germany, Japan and Korea have approximately 100 people per kilometre. Several countries (e.g. Australia, Canada, Scotland) distinguish between "rural" and "remote" in terms of workforce availability and health service provision. While rural and remote areas have some common issues; there are also significant differences (e.g. the absence of resident medical practitioners in remote areas). In Australia the core remote health workforce consists of remote area nurses (RANs) and Aboriginal and Torres Strait Islander health workers, with support from a range of multi-disciplinary health professionals across allied health and medical professions (Health Workforce Australia 2013). Whereas in Scotland, 'rural' and 'remote' includes those living in the mainland highlands, where transportation is limited and small populations living on islands that are only accessible by boat or plane (Ono et al. 2014). Similarly, Japan uses 'isolated rural areas' (called hekichi) to describe the terrain (e.g. islands) and climate (e.g. mountainous area with heavy snow) that hinder access to health services (Ono et al. 2014).

The research studies described in this book were conducted in Australia and used the Remoteness Areas (RAs) determined by the Australian Bureau of Statistics (ABS) in the Accessibility Remoteness Index Australia (ARIA) (ABS 2006). The ABS (2003) developed these categories to establish common terminology for data analysis examining what was often referred to as 'urban', 'rural', 'remote', 'metropolitan', 'regional', 'the bush' or 'the outback'. They concluded that the critical concept was 'remoteness' and suggested that what defined 'city' and 'country' in this context was how far someone travels to access infrastructure, thus RAs are measures of 'remoteness of a point based on the physical road distance to the nearest Urban Centre' (Australian Bureau of Statistics [ABS] 2006, p. 40). The ABS categorise Australia into five geographic regions: Major cities; Inner Regional; Outer Regional; Remote; and Very Remote. Approximately 1% of the Australian population live in very remote areas and approximately 2% live in remote areas (ABS 2003). For this book, the participants in the study are described as working in 'remote' areas if they work in a remote or a very remote region as categorised by the RAs defined by the ABS.

The Distinction Between Rural and Remote

The distinction between rural and remote is contentious and differs depending on perspective. For example, health professionals, teachers, statisticians, councils and governments all have different ways of distinguishing geographical boundaries. In

Australia, a country with a large land mass and a relatively small population. Most of the Australian population live near the coast in cities, and a small proportion of the population live in regional areas that are largely urbanised and provide infrastructure and services comparable to those found in cities. An even smaller proportion of Australia's population live in remote areas that are not only great distances from cities and regional centres, they have limited access to the infrastructure and services that the city and regional populations often take for granted. In Australia, some of the towns that are classified as remote in the ABS system, may be considered rural by other systems of categorisation. To determine whether this distinction would affect the ideas presented in this book a systematic scoping literature review was conducted. The review found that the challenges that health professionals experience in remote and very remote areas are more similar than they are different so collapsing the remote and very remote categories into one category called 'remote' was not believed to have negatively impacted the study's findings (Onnis and Pryce 2016). However, it is acknowledged that remote areas also comprise areas that within a different system of categorisation may be considered rural. The different interpretations of 'remote', particularly the differentiation between remote and rural is acknowledged, and therefore in the interest of transparency this chapter is clearly identifying that assumptions were made based on the available evidence at hand.

International Remote Contexts

'The nursing education programs, healthcare systems and Aboriginal cultures in Canada and Australia are surprisingly alike despite geographical distances, climatic extremes, and differences in government policies.' (Kent-Wilson 2010, p. 59)

An examination of the comparability of Australia's remoteness classifications with those of other geographically large countries suggests that they are comparable, provided that context is considered before drawing conclusions (Dieleman et al. 2011). In their study about rural-urban disparities, Pong et al. (2009, p. 58) used Canada's Metropolitan Influence Zone classification and the Accessibility/Remoteness Index of Australia (ARIA), as both indexes use access to infrastructure to determine remoteness. While Australia used distance by road, Canada used travel costs, with the primary transport methods being by road and for communities that were not connected to the main road network, travel cost were calculated from the least expensive travel options available (Alasia et al. 2017). Further, a review of the World Health Organisation website reveals that remoteness is most commonly determined by distance—be it time travelled, or cost of travel to infrastructure and services (Alasia et al. 2017; Ameli and Newbrander 2008). Therefore, the content contained in this book will be relevant to managers of health professionals in rural and remote areas, beyond northern tropical Australia as there are comparable understandings of geographical remoteness.

Ultimately, effective management practices contribute to improving access to health services and the health of people living in remote regions. This serves as a reminder that management practices influence the implementation and localisation of HRM policies and HR outcomes in *all* organisations. However, context is important, which makes this book relevant to managers in comparable remote working conditions to those described throughout this book regardless of the country in which they work. That is, remote managers in Australia will have more in common with remote managers in Canada, Scotland and the United States, than they will with city-based managers in Australia.

How to Use to This Book

The book is structured so that each chapter builds on the previous chapter. Throughout the book there are opportunities to reflect and consolidate ideas being generated about how to localised your management practices. There are also pointers to further resources, items you may wish to add to your toolkit, and examples of the ideas in practice.

⑪ Reflection
The sections marked 'Reflection' are opportunities to pause and reflect on yourself, your workplace, your management style and your management practices. It is recommended that you keep a reflective journal to capture your reflections while reading this book. These reflections will shape your management practices, and will influence how you manage your current and future teams.

⊤ Resources
There are signposts throughout the book directing you towards additional resources. These resources include further reading, helpful websites, online information and management tools.

⊞ Manager's Toolkit
You may have heard the saying 'when all you have is a hammer, every problem looks like a nail.' This is true for many managers who find themselves managing every situation in the same way; even though they know that not all situations require the same approach. Why do they do this? Often because they are so busy it is their first reaction; however, sometimes it is because they are not aware of the other options available to them. For this reason, as you read through the book you will be able to collect ideas about management practices that will suit different situations, including tools and resources. These are tools that you think will be helpful to you, as a manager. By the end of the book, your Manager's toolkit will be brimming with management tools, so that when you are faced with a 'people management' issue, you can select the appropriate tool to suit the situation. Because sometimes, no matter how hard you try a hammer is not going to solve a problem that needed jump leads!

In Practice

The 'In Practice' sections are stories from people working in remote areas about their experiences. Some of the human resource management (HRM) concepts discussed in this book look different when implemented in different settings, so the 'In Practice' examples are presented to show the range of ways a concept can be localised to your remote location. It links the theory and discussion to practice.

Why Does Workforce Sustainability Matter?

The World Health Organisation (WHO) predicts global health workforce shortages with some regions already experiencing shortages of nurses (Campbell et al. 2013; WHO 2010). These shortages arise from low numbers of trained health professionals in some countries, and efforts to increase the number of health professionals being trained does not always translate into an increased number in the workforce. For example, some countries are bearing the burden of training medical staff who once qualified, move to countries where health professionals are better compensated, where career pathways exist and the working conditions are better. In most cases, rural and remote communities are not high on the desirable work location list. Hence, health service providers in remote regions worldwide, traditionally face challenges in attracting and retaining health professionals (Dolea et al. 2010; Mbemba et al. 2013).

To understand the challenges of turnover some examples are offered. In Australia's Northern Territory the turnover rate for Remote Area Nurses 'varied from about 30% to nearly 90% over a three year period' (Garnett et al. 2008, p. 31). More recently, Russell et al. (2017) found that for nurses and allied health professionals in Australia's Northern Territory turnover was extremely high, with average annual turnover rates being 66% (no longer working in any remote health centre) and 128% (no longer working in that specific remote health centre). This means that on average two-thirds of the staff leave not only the health service but they do not work in any remote health service 12 months later. The second turnover rate shows some mobility between health services, but it is concerning as a 128% suggests that there is more than a complete changeover of staff. Furthermore, Russell et al. 2017 found low twelve month stability rates (20%) with half leaving in the first four months of employment. Similarly, an examination of the General Practitioner (GP) mobility data from 2008 to 2012 showed a higher mobility rate for GPs moving *from* remote regions than moving *into* remote regions (McGrail and Humphreys 2015). This means that each year the number of people going to remote areas is less than the number of people leaving which has ongoing implications for access to health services. This type of workforce instability in rural and remote areas is observed in many countries around the world (Hayes et al. 2006).

In fact, global health workforce shortages including the additional challenges for remote regions are recognised. Despite the known challenges, there are remote regions where the health services have positive outcomes, and remote managers who are thriving in the challenging circumstances. Hence, there are areas of relative stability within a sector of high turnover and with this comes an opportunity to learn about how workforce sustainability can be achieved. The following chapters contain stories adapted from real life accounts of working in remote settings; tales of triumph in adversity and narratives about the resilience of the people who work in remote regions. Whether they are romanticised recollections or matter-of-fact accounts, they paint a picture about the extreme, adverse and rewarding conditions in which remote health professionals work.

In responding to the challenges of high turnover, a variety of flexible service delivery methods have been implemented including fly-in/fly-out (FIFO) or drive-in/drive-out (DIDO) models, a combination of face-to-face interaction and technology-based interactions (e.g. telemedicine), outreach services, hub and spoke models, agency staff and locums, and international recruitment programs (Margolis 2012). As well as these flexible models of service provision there are traditional models that include sole practitioners (private and/or public), and community-based models. The salient message here is that there are a variety of flexible service models created to respond to the demand for health services so that remote populations can benefit from healthcare regardless of their geographic location.

There is no easy solution, health is an expensive business and human life is priceless. Health service managers are tasked with balancing this cost equation daily. For example, they must decide whether to bring the patient to the health service or take the health service to the patient. Economies of scale when working with life and death are complicated. For remote residents, travel to major cities for health treatment can be expensive, traumatic and introduce further challenges for them, their family and local communities. In addition, there are many challenges for health professionals living and/or working in remote regions and for organisations providing health services who seek competent health professionals for regions where recruitment is difficult and turnover high, all of which impact on the continuity of patient care (Productivity Commission 2005; Russell et al. 2017).

Globally, health workforce shortages are reported and the additional challenges for remote regions are recognised. Therefore, the influence of sustainable workforces in remote regions has an impact not only on the health services delivered but on the health outcomes of remote populations. Several reports have shown that there are differences in the health of populations living in remote regions when compared to the health of people living in cities, with poorer health outcomes associated with the further away people live from major cities (AIHW 2012; WHO 2010). People who live in remote regions are aware of the distance that they live from cities and usually have realistic expectations about what is possible for the location in which they live. Travelling great distances, on unmade roads, through extreme weather, and unanticipated natural obstacles, are common occurrences for residents of remote regions worldwide. For people living in remote areas accessing health services is often made more difficult where there is frequent turnover of health professionals or where there are long term vacancies in key positions (e.g. GP, midwife). Despite the known challenges, there are remote regions where the health services have positive outcomes, where health workforces are sustainable, and above all remote managers are thriving in the challenging circumstances.

☝ Resources

For remote managers it is important to understand turnover, stability and mobility data.

Turnover: is the rate at which employees leave the workforce and are replaced by new employees. Turnover is a natural part of the employment cycle; and is not a negative thing in itself. Turnover brings new people with new ideas into the organisations. However, turnover can be a workplace issue if there is a high rate of voluntary turnover, particularly if new employees leave prematurely (e.g. within a few weeks or months of commencing). Low turnover can also be an issue for organisations because it limits the opportunity to bring new skills into the organisation and it can be detrimental where there are small workforces, such as remote health centres, where some employees stay for too long which can be difficult to manage, and is often an unhealthy situation for the employee as well (i.e. it would be in the best interest of the person and the organisation for them to leave).

Turnover is categorised as voluntary or involuntary. Involuntary turnover is initiated by the organisation (e.g. retrenchment, redundancy, short-term fixed contracts). Voluntary turnover is initiated by the employee (e.g. people leave a job to start another job, for personal reasons, career change). Voluntary turnover includes resignations from people who do not have another job to go to and are often a reflection of poor recruitment. Recruitment will be discussed in more detail in Chap. 4.

To calculate you annual turnover:

$$\text{Annual Turnover Rate} \quad = \quad \frac{\text{Number of employee that left}}{\text{Average Number of Employees*}} \quad \times \ 100$$

*Average number of employees is (number at the start) + (number 12 months later) divided by 2

Further information about calculating employee turnover is available: https://resources.workable.com/tutorial/calculate-employee-turnover-rate.

Workforce stability: The term workforce stability is used to describe those people who stay. Stability is generally regarded as being the opposite of workforce turnover (Buchan 2010). One indicator used to calculate workforce stability is the average years in a position by work group (e.g. nurses, department) or location (e.g. community or remote health centre). Another measure of stability is to calculate the stability rate for each work group or location. The stability rate is the proportion of employees who were in a position at the beginning of the year who are still in the same position at the end of the year. This could be calculated over 12 months or a longer period (e.g. 2 years, 5 years). Calculating stability rates can help managers to see where there is stability rather than just focusing on turnover. Stability is more closely associated with retention than turnover because understanding stability rates and why people stay can help managers to improve retention rather than focus on those they left the organisation (Buchan 2010).

Labour mobility: Labour mobility is the ability of people to move between jobs. It is an important aspect of workforce flexibility because movement within the labour market allows people to be matched with a suitable job that fits their preferences and in which they are productive. Current data about labour mobility can be found on the Australian Bureau of Statistics website. Data, including trends, specific to your region or profession may also be available (e.g. Medicine in Australia: Balancing Employment and Life (MABEL)).

 **Resources**
Metrics for determining turnover and retention rates

In the article, *How best to measure health workforce turnover and retention: Five key metrics,* Russell et al. (2012) methodologically illustrate the use of five key workforce turnover and retention metrics suitable for use by Australian rural health services. These metrics are: crude turnover (separation) rates, stability rates, survival probabilities, median survival and Cox proportional hazard ratios. The article includes information, including examples about how they are calculated using data from Australian rural and remote health services. This article is a good guide for managers wishing to understand what these metrics mean and how they can be calculated.

Manager's Toolkit
You can start to build your toolkit with information to help you to do workforce calculations such as turnover. It is helpful to know the average for your industry. To understand the turnover rate and the stability rates, and if there are patterns, e.g. are there particular positions that turnover more frequently than others? This will help you to understand why you are seeing the turnover that you are seeing, is there something about the position?... The person?... The selection process?

What Does the Human Resource Management (HRM) Approach Offer?

The HRM approach offers an alternative, yet complementary perspective for examining health workforce sustainability in remote regions. HRM describes the management of people within the parameters of an employment relationship. HRM is: a set of strategies and practices to gain commitment and loyalty; a function that creates competence and feelings of belongingness in an organisation; and an approach consisting of policies, practices and systems that influence a person's behaviour, attitude and performance at work (De Cieri et al. 2004; Thompson 2011). A more holistic view, and the one adopted for this book, builds on the work of Beer et al. (1984) who described HRM as the development of everything within an organisation that supports effective management practices. It is envisaged that remote health managers will see value in HRM once they understand that HRM practices are there to assist them to undertake their management role both effectively and efficiently (Kabene et al. 2006).

In remote regions, work and home life interconnect, workers are immersed in new cultures, and social support networks are crucial; all indicating that HRM has much to offer remote managers. In remote regions worldwide, where personal and professional lives co-exist, an employee's relationship with their immediate manager may significantly influence workforce sustainability. Furthermore, social isolation is reported as a challenge when working in remote regions; therefore, socialisation into the organisation through orientation and onboarding strategies, may be especially important, particularly in terms of social-identity. It is proposed that where people develop a social-identity with the organisation for which they work, it influences their organisational commitment and shapes psychological links formed between the organisation and the employee (Gould-Williams and Davies 2005; Highhouse et al. 2007). These psychological links (which develop through psychological contracts) arise from an employee's beliefs about the employment relationship and the obligations of their employer (Knights and Kennedy 2005). The influence that these psychological contracts have on workforce sustainability is explored in greater detail in chapter four.

Despite the implementation of flexible models of service, rewards and financial incentives, and socialisation of remote employees; workforce sustainability challenges continue for many organisations providing quality health services in remote regions (Chisholm et al. 2011; Garnett et al. 2008; Hunter et al. 2013; Russell et al. 2017; Weymouth et al. 2007). HRM practices offer solutions for workforce challenges particularly with regard to overcoming workforce shortages where retention is central to ensuring 'locally delivered, appropriate and sustainable health services' (Chisholm et al. 2011, p. 87). HRM, which focuses on managing employment relationships, offers a solid evidence-base, including theoretical frameworks, to examine workforce issues (Safdar 2012).

The health sector more frequently focuses on healthcare-related competence and the challenges of remoteness, and is generally non-cognisant of the benefits of an HRM approach. Therefore, applying evidence-based HRM theories and concepts provides both a richer and deeper understanding about the benefits of incentives, development programs, and reward systems that are relevant to the workforce sustainability challenges reported by remote health professionals and remote managers. Hence, an examination of the challenges from the perspectives of managers and health professionals who are currently working in remote regions, using an HRM approach, offers a solutions-based approach rather than the deficit approach which is often used when discussing turnover (Gorton 2015; Wakerman and Humphreys 2012). The solutions-focused HRM approach shaped the development and content of this book.

References

Alasia A, Bédard F, Bélanger J, Guimond E, Penney C (2017) Measuring remoteness and accessibility—a set of indices for Canadian communities, Catalogue no. 18-001-X

Ameli O, Newbrander W (2008) Contracting for health services: effects of utilization and quality on the costs of the basic package of health services in Afghanistan. Bull World Health Organ 86:920–928

Australian Bureau of Statistics (ABS) (2003) ASGC Remoteness classification: purpose and use (Census Paper No. 03/01). Canberra, Australia. http://www.abs.gov.au. Accessed 2 June 2018

Australian Bureau of Statistics (ABS) (2006) Australian standard geographical classification (ASGC). ABS, Canberra, Australia. http://www.abs.gov.au. Accessed 2 June 2018

Australian Human Rights Commission (2005) Questions and answers about aboriginal & Torres Strait islander peoples. http://www.humanrights.gov.au/publications/questions-and-answers-about-aboriginal-torres-strait-islander-peoples. Accessed 12 Oct 2018

Australian Institute of Health and Welfare (AIHW) (2012) Australia's health 2012. Australia's health series (No.13. Cat. no. AUS 156). AIHW, Canberra, Australia. http://www.aihw.gov.au/publication-detail/?id=10737422172. Accessed 5 Sept 2017

Beer M, Spector B, Lawrence PR, Mills DQ, Walton RE (1984) Managing human assets. The Free Press, New York

Bent A (1999) Allied health in Central Australia: challenges and rewards in remote area practice. Aust J Physiother 45(3):203–212

Buchan J (2010) Reviewing the benefits of health workforce stability. Hum Resour Health 8(29)

Campbell J, Dussault G, Buchan J, Pozo-Martin F, Guerra Arias M, Leone C, Siyam A, Cometto G (2013) A universal truth: no health without a workforce. WHO Press, Geneva http://www.who.int/workforcealliance/knowledge/resources/GHWA_AUniversalTruthReport.pdf. Accessed 3 Sept 2017

Chisholm M, Russell D, Humphreys J (2011) Measuring rural allied health workforce turnover and retention: what are the patterns, determinants and costs? Aust J Rural Health 19(2):81–88

De Cieri H, Kramar R, Noe R, Hollenbeck J, Gerhart B, Wright PM (2004) Human Resource Management in Australia: strategy, people, performance.McGraw Hill Australia Pty Ltd., Macquarie Park, Australia

Dieleman M, Kane S, Zwanikken P, Gerretsen B (2011) Realist review and synthesis of retention studies for health workers in rural and remote areas, World Health Organization. http://apps.who.int/iris/bitstream/10665/44548/1/9789241501262_eng.pdf. Accessed 2 June 2018

Dolea C, Stormont L, Braichet J (2010) Evaluated strategies to increase attraction and retention of health workers in remote and rural areas. Bull World Health Organ 88:379–385

Garnett S, Coe K, Golebowska K, Walsh H, Zander K, Guthridge S, Li S, Malyon R (2008) Attracting and keeping nursing professionals in an environment of chronic labour shortage: a study of mobility among nurses and midwives in the northern territory of Australia. Charles Darwin University Press, Darwin. http://digitallibrary.health.nt.gov.au/dspace/bitstream/10137/228/1/nurse_report.pdf. Accessed 5 Sept 2017

Gorton S (2015) Who paints the picture? Images of health professions in rural and remote student resources. Rural Remote Health 15(3):3423

Gould-Williams J, Davies F (2005) Using social exchange theory to predict the effects of HRM practice on employee outcomes. Public Manag Rev 7(1):1–24

Hawthorne L (2001) The globalisation of the nursing workforce: barriers confronting overseas qualified nurses in Australia. Nurs Inquiry 8(4):213–229

Hayes L, O'Brien-Pallas L, Duffield C, Shamian J, Buchan J, Hughes F, Spence Laschinger HK, North N, Stone PW (2006) Nurse turnover: a literature review. Int J Nurs Stud 43(2):237–263

Health Workforce Australia (HWA) (2012) Australia's health workforce series—doctors in focus Adelaide. HWA, Australia

Health Workforce Australia (2013) National rural and remote workforce innovation and reform strategy. HWA, Adelaide

Health Workforce Australia (HWA) (2014) Health workforce by numbers—Issue 3. HWA, Australia

Hegney D, McCarthy A, Rogers-Clark C, Gorman D (2002a) Retaining rural and remote area nurses. The Queensland, Australia experience. J Nurs Adm 32(3):128–135

Hegney D, McCarthy A, Rogers-Clark C, Gorman D (2002b) Why nurses are attracted to rural and remote practice. Aust J Rural Health 10(3):178–186

Hegney D, McCarthy A, Rogers-Clark C, Gorman D (2002c) Why nurses are resigning from rural and remote Queensland health facilities. Collegian 9(2):33–39

Highhouse S, Thornbury EE, Little IS (2007) Social-identity functions of attraction to organizations. Organ Behav Hum Decis Process 103(1):134–146

Hunter E, Onnis L, Santhanam-Martin R, Skalicky J, Gynther B, Dyer G (2013) Beasts of burden or organised cooperation: the story of a mental health team in remote Indigenous Australia. Australas Psychiatry 21(6):572–577

Kabene SM, Orchard C, Howard JM, Soriano MA, Leduc R (2006) The importance of human resources management in health care: a global context. Hum Resour Health 4(20)

Kent-Wilkinson A, Starr L, Dumanski S, Fleck J, LeFebvre A, Child A (2010) International nursing student exchange: rural and remote clinical experiences in Australia. J Agromedicine 15(1):58–65

Knights JA, Kennedy BJ (2005) Psychological contract violation: impacts on job satisfaction and organizational commitment among Australian senior public servants. Appl HRM Res 10 (2):57–72

Lippel K, Johnstone R, Baril-Gingras G (2017) Regulation, change and the work environment. Relations Industrielles/Ind Relat 72(1):3–16

MacLeod Martha L. P, Stewart Norma J, Kulig Judith C, Anguish P, Andrews M. E, Banner D, Garraway L, Hanlon N, Karunanayake C, Kilpatrick K, Koren I, Kosteniuk J, Martin-Misener R, Mix N, Moffitt P, Olynick J, Penz K, Sluggett L, Van Pelt L, Wilson E, Zimmer L (2017) Nurses who work in rural and remote communities in Canada: a national survey. Hum Resour Health 15(1):34

Margolis S (2012) Is fly in/fly out (FIFO) a viable interim solution to address remote medical workforce shortages? Rural Remote Health 12:2261

Mbemba G, Gagnon M, Paré G, Côté J (2013) Interventions for supporting nurse retention in rural and remote areas: an umbrella review. Hum Resour Health 11(44)

McGrail MR, Humphreys JS (2015) Geographical mobility of general practitioners in rural Australia. Med J Aust 203(2):92–96

Newhook J, Neis B, Jackson L, Roseman S, Romanow P, Vincent C (2011) Employment-related mobility and the health of workers, families, and communities: the Canadian context. Labour (Spring):121–156

Onnis L, Pryce J (2016) Health professionals working in remote Australia: a review of the literature. Asia Pac J Hum Resour 54:32–56

Ono T, Schoenstein M, Buchan J (2014) Geographic imbalances in doctor supply and policy responses, OECD Health Working Papers, No. 69. OECD Publishing, Paris. http://dx.doi.org/ 10.1787/5jz5sq5ls1wl-en. Accessed 2 June 2018

Pong R, Desmeules M, Lagace C (2009) Rural-urban disparities in health: how does Canada fare and how does Canada compare with Australia? Aust J Rural Health 17(1):58–64

Productivity Commission (2005) Australia's health workforce, Research Report. http://www.pc. gov.au/inquiries/completed/health-workforce/report/healthworkforce.pdf. Accessed 5 Sept 2017

Robertson R (2014) Leading on the edge. Wiley, Australia

Russell D, Humphreys J, Wakerman J (2012) How best to measure health workforce turnover and retention: five key metrics. Aust Health Rev 36(3):290–295

Russell DJ, McGrail MR, Humphreys JS (2017) Determinants of rural Australian primary health care worker retention: a synthesis of key evidence and implications for policymaking. Aust J Rural Health 25(1):5–14

Safdar R (2012) Relative and cross-national human resources management research: development of a hypothetical model. Glob J Manag Bus Res 12(2):6–19

Thompson P (2011) The trouble with HRM. Hum Resour Manag J 21(4):355–367

Torres Strait Islander Regional Council (2016) Our geography. http://www.tsirc.qld.gov.au/. Accessed 29 Sept 2017

Wakerman J, Humphreys J (2012) Sustainable workforce and sustainable health systems for rural and remote Australia. Med J Aust 199(5):14–17

Weymouth S, Davey C, Wright J, Nieuwoudt L, Barclay L, Belton S, Svenson S, Bowell L (2007) What are the effects of distance management on the retention of remote area nurses in Australia? Rural Remote Health 7(3):652

World Health Organisation (WHO) (2010) Increasing access to health workers in remote and rural areas through improved retention. WHO Press, France. http://www.searo.who.int/nepal/mediacentre/2010_increasing_access_to_health_workers_in_remote_and_rural_areas.pdf. Accessed 5 Sept 2017

Zuber-Skerritt O, Wood L, Louw I (2015) A participatory paradigm for an engaged scholarship in higher education. Sense Publishers, Rotterdam

Remote Workplaces

2

> *People leave managers, not companies. So much money has been thrown at the challenge of keeping good people– in the form of better pay, better perks, and better training– when in the end, turnover is mostly a manager issue.*
> Marcus Buckingham and Curt Coffman, The Gallup Organisation.
> (Buckingham and Coffman 1999, p. 27)

Key Messages

- The 'remote workplace' is the 'space' in which there is the most opportunity for managers to influence decision-making about turnover and retention.
- If health professionals make decisions about whether to remain or leave based on their experience in the 'remote workplace' it follows that the manager, the most influential person in that workplace, will have a significant impact on turnover and retention.
- The changing structure of contemporary workforces and the mobility of individual employees, create an opportunity for managers to improve retention through creative workplace solutions.
- In a competitive labour market, management practices may be the competitive advantage needed to improve retention, particularly in remote regions where demand exceeds supply.

'Nurses will give up their anonymity when they go bush. There will be little personal privacy. They will not be able to 'go home' from their job. Home will be just a stroll or short drive from work. People will knock on their door in the middle of the night to seek help. They will be asked to do consults in the supermarket or while socialising. Many will feel they are living their life in a 'fish bowl' ... Many of the challenges associated with remote practice such as those touched on above cannot be avoided. They are part and parcel of remote work. They are not necessarily good or bad – it is just the way it is.' (Kelly 2000)

© Springer Nature Singapore Pte Ltd. 2019
L. Onnis, *HRM and Remote Health Workforce Sustainability*,
Management for Professionals, https://doi.org/10.1007/978-981-13-2059-0_2

For many remote managers it can feel like you are the only person facing the challenges that you have daily. However, the literature describes common challenges and rewards suggesting that it is unlikely that you are the only person facing these challenges. In fact, the common challenges suggest that most remote managers have similar experiences. While this is reassuring, it also serves to highlight areas where improvement efforts will most likely have the greatest impact on workforce sustainability.

There is extensive literature about the rewards and challenges of working in remote regions, particularly, high turnover and workforce retention. Onnis and Pryce (2016, p. 49) seeking to understand the challenges reported by health professionals working in remote regions, found that 'the reasons that influence a health professional's decision to remain or leave are not only diverse, they are inconsistent, that is, one health professional's reason for leaving may be another one's reason for staying'. Furthermore, the attraction to working in remote regions is varied (McGrail et al. 2011) and regardless of the health professional's motivation, expectations about both the professional role and the organisation, influence job satisfaction and ultimately turnover (Knights and Kennedy 2005).

A review of the challenges and rewards reported in the literature, highlighted areas where there are opportunities to improve the working experience for health professionals working in remote regions. Therefore, it is anticipated that many of the challenges described in this chapter can be observed in remote regions globally. While the majority of studies reported in the literature reviewed were Australian, international studies were not excluded. This suggests that the international evidence is limited in this area; however, the findings were similar suggesting that it is reasonable to propose that the findings discussed in this book are relevant to managers working in remote regions other than tropical northern Australia (Belaid et al. 2017). The review of the literature in this area suggested that addressing these challenges will improve workforce sustainability (Devine 2006; Greenwood and Cheers 2002; Hays et al. 2003; O'Toole and Schoo 2010; Wakerman et al. 2009). The interrelatedness of these challenges and more specifically their association with workforce sustainability are discussed in this chapter.

Synthesis of the Known Rewards and Challenges

This section uses a synthesis of the literature to provide a broad overview of the most frequently identified rewards and challenges providing an evidence-based summary of what is currently known. The references are also provided for remote managers wishing to read further on the specific rewards and challenges identified. The first challenge discussed, person-fit, is central to the notion of having the right people, in the right place, with the right skills.

Person-Fit (Personal)

- Person-fit is influenced by personal characteristics, such as: age, rural upbringing, previous exposure to living in a rural or remote area (e.g. first qualification, clinical placements) (Hegney et al. 2002a, b, c; Kruger and Tennant 2005; Kent-Wilkinson et al. 2010).
- For effective remote practice, health professionals need to be self-reliant, have highly developed problem-solving skills, be flexible, be responsive to the environment, organised, and creative so that they can work with limited resources, and be able to cope with adversity (Bent 1999; Devine 2006; Hays et al. 2003; Hegney et al. 2002a).

Health and Wellbeing (Personal)

- Social and emotional challenges exist, with family and friends being both supportive (e.g. social networks reduce feelings of isolation and loneliness) and the cause of conflict between obligations and responsibilities (e.g. role-conflict where friends from the local community are also clients; Indigenous Health Workers who also have community responsibilities) (Birks et al. 2010; Hays et al. 2003; Hegney et al. 2002a, b, c; Greenwood and Cheers 2002; Kruger and Tennant 2005; Kent-Wilkinson et al. 2010; O'Toole and Schoo 2010; Bent 1999; Hays et al. 2003).
- A lack of anonymity exists where professional and social lives merge (Gardiner et al. 2005).
- The challenges of balancing work and family responsibilities, such as: workloads impeding on family responsibilities, reduced access to childcare, inadequate housing, the high cost of living, limited employment opportunities for partners, and poor educational options (Bent 1999; Gardiner et al. 2005; Hays et al. 2003; Hegney et al. 2002a, b, c; Humphreys et al. 2002; Kent-Wilkinson et al. 2010; Kruger and Tennant 2005; Opie et al. 2011; Wakerman et al. 2009).
- The health implications differ between people, which are often due to each individual's personal coping strategies, and self-care regimes. Frequently reported health implications include: stress, distress, feeling exhausted, overwhelmed, tired and fatigued (Birks et al. 2010; Devine 2006; Gardiner et al. 2005; Greenwood and Cheers 2002; Hays et al. 2003; Kruger and Tennant 2005; Lenthall et al. 2011; Opie et al. 2011).
- The personal health consequences: bullying, injury and trauma (Jackson et al. 2002).
- Even when health professionals recognise the need for a break, there can be challenges in accessing annual leave due to workforce shortages and excessive on-call responsibilities (Birks et al. 2010; Hays et al. 2003; Humphreys et al. 2002; Kruger and Tennant 2005; Opie et al. 2011).

Differences Between Professions (Professional)

- There are differences in the challenges described by each profession (e.g. nursing, allied health).
- Some private practitioners reported challenges in working across both private and public systems to sustain themselves financially, and the challenges of practicing in a community with a limited client base (Greenwood and Cheers 2002; O'Toole and Schoo 2010).
- Some reported challenges with large clinical caseloads; but, spoke positively about the diversity of caseloads (Birks et al. 2010; Devine 2006; Hegney et al. 2002a, b; Kruger and Tennant 2005).
- The diversity and variety of the work in general encouraged many to remain because they enjoyed the professional challenges and opportunities (Bent 1999; Devine 2006; Greenwood and Cheers 2002; Hays et al. 2003; Hegney et al. 2002a, b, c; Humphreys et al. 2002; Kruger and Tennant 2005; Opie et al. 2011; Wakerman et al. 2009).
- For some health professionals their training prepared them for remote practice; however, for others there was limited understanding about remote practice and a lack of clinical support within their discipline (Battye and McTaggart 2003; Devine 2006; Greenwood and Cheers 2002).
- There is inequity in the salary and benefits paid to allied health professionals, health workers, doctors and nurses (e.g. the financial benefits received through incentive such as the Remote Area Nurses Incentive Package (RANIP), access to housing, etc.) (Santhanam et al. 2006).

Professional Development (Professional)

- There are challenges and difficulties accessing professional development (Bent 1999; Birks et al. 2010; Devine 2006; Hegney et al. 2002a, b; Lenthall et al. 2011; O'Toole and Schoo 2010). For many health professionals, access to education and training is crucial for maintaining their skills, particularly with the additional education needed for advanced nursing roles (Bent 1999; Hays et al. 2003; Hegney et al. 2002a, b; Lenthall et al. 2011).
- There are perceived impediments for accessing professional development including: cost, travel time, and the need for backfill or a locum to cover the clinical role (Bent 1999; Battye and McTaggart 2003; Hegney et al. 2002a, b).
- The challenges associated with career progression, including limited remote career paths which are disincentives for health professionals wanting to move into senior positions (Bent 1999; Devine 2006; Hegney et al. 2002a, c; O'Toole and Schoo 2010; Wakerman et al. 2009).
- The challenges in accessing professional support (e.g. supervision, mentoring, professional and peer support networks) which were influenced by location, individual characteristics, and the model of service delivery within which they

worked (Battye and McTaggart 2003; Bent 1999; Birks et al. 2010; Devine 2006; Gardiner et al. 2005; Greenwood and Cheers 2002; Hays et al. 2003; Hegney et al. 2002b, c; Santhanam et al. 2006; Wakerman et al. 2009).

- Professional development highlighted the clinical skills needed to work in remote areas; however, education around challenges such as managing stress, and personal coping skills were considered as beneficial as clinical education (Gardiner et al. 2005).

HRM Policy Choices (Organisational)

- HRM policies both hinder and support the individual health professional's experience. These include, but are not limited to: access to paid professional development, study leave, conference attendance, reimbursement or subsidies for relocation, access to annual leave, travel subsidies, childcare subsidies, inadequate, poor or inappropriate housing, salary packaging, financial consideration for being on-call 24 hours a day and remuneration (Battye and McTaggart 2003; Birks et al. 2010; Devine 2006; Hegney et al. 2002a, b, c; Kruger and Tennant 2005; O'Toole and Schoo 2010; Wakerman et al. 2009).

Management Practices (Organisational)

- Health professionals described a general dissatisfaction with management practices. The challenges were a result of a lack of management support, inappropriate line supervision and lack of management recognition of their work (Battye and McTaggart 2003; Gardiner et al. 2005; Hegney et al. 2002a, b; Opie et al. 2011).
- The specific challenges with management practices included: unrealistic expectations of staff, poor co-ordination of visiting services, poor resourcing, difficulties accessing transport and overnight accommodation, long working hours, an excessive amount of unpaid work and administrative tasks, and poorly arranged backfill and delays in filling vacancies (Battye and McTaggart 2003; Bent 1999; Birks et al. 2010; Devine 2006; Hays et al. 2003; Hegney et al. 2002b; Lenthall et al. 2011; Opie et al. 2011; O'Toole and Schoo 2010; Santhanam et al. 2006).

Environment, Climate and Culture (Contextual)

- Remote regions hold many challenges for the local population that are also experienced by remote health professionals, most noticeably service delivery to large geographical areas which, involves travelling long distances with reduced clinical working time (Battye and McTaggart 2003; Bent 1999; Devine 2006). The environment increases the hazards and can require air travel and roads are often roads unreliable and impassable (Alasia et al. 2017; Battye and McTaggart 2003).

- Health professionals working in remote regions need to develop the necessary skills to be effective in remote areas as they are responsible for providing healthcare to a range of clients, often with their own first languages (Battye and McTaggart 2003; Bent 1999; Kent-Wilkinson et al. 2010).
- Health professionals work in culturally diverse communities and must show a high degree of cultural awareness, cultural knowledge and provide culturally appropriate congruent care (Battye and McTaggart 2003; Bent 1999; Kent-Wilkinson et al. 2010; Santhanam et al. 2006). This may include adopting a community focus rather than a medical focus, and building community relationships (Greenwood and Cheers 2002; Devine 2006; Santhanam et al. 2006).
- For some health professionals the opportunity to work with Indigenous people is the attraction of remote work and they adapt to provide cross-cultural services using flexible models of service delivery addressing the needs of clients in unique rural and remote environments (Bent 1999; Greenwood and Cheers 2002; Humphreys et al. 2002; Kent-Wilkinson et al. 2010).

Rewards

- Many health professionals found their job satisfying, saying that they, enjoyed the lifestyle, climate, and opportunities to travel; felt valued by their employer; had family friendly workplaces; and their partners were engaged with community activities and/or had found satisfying work locally (Bent 1999; Devine 2006; Gardiner et al. 2005; Greenwood and Cheers 2002; Hays et al. 2003; Hegney et al. 2002a, b, c; Kruger and Tennant 2005; Opie et al. 2011).
- The more experienced health professionals enjoyed the autonomy, where independence increased professional responsibilities (Bent 1999; Devine 2006; Hays et al. 2003; Hegney et al. 2002a, b, c; Kruger and Tennant 2005; Opie et al. 2011; Santhanam et al. 2006).
- Some other rewarding aspects of remote work included feeling valued and respected by the community (Devine 2006; Hays et al. 2003; Hegney et al. 2002b, c).

⫟ Resources

Further information about the literature review and all of the findings can be found in the article, *Health professionals working in remote Australia: a review of the literature* (Onnis and Pryce 2016).

Available from: https://researchonline.jcu.edu.au/41291/.

What Can We Learn from the Synthesis of the Literature?

The synthesis of the literature highlighted the key challenges and rewards reported by remote health professionals. Remoteness is often regarded as a challenge and said to be a key factor in turnover and retention. While the role of remoteness is not disputed, health professional's more frequently report other factors that influence their decisions. There is a sense that most health professionals can learn to accept the context and the environment (e.g. geography, culture, and climate) as challenges; however, they report frustration with systems and workplaces that do not support them personally and professionally (Battye and McTaggart 2003; Hegney et al. 2002a). In remote regions, where work and personal lives co-exist, an employee's relationship with their immediate line manager and colleagues, are not only important to the professional working relationship, they influence decisions that often result in voluntary turnover.

In an endeavour to understand the challenges that health professionals working in remote regions were reporting, Onnis and Pryce (2016) identified four themes: Personal, Professional. Organisational and Contextual. Some of these themes have been discussed in other forums by other researchers and health sector managers and will undoubtedly be familiar to remote health professionals and managers. However, it is not the identification of these themes that helps us to understand the factors that influence turnover and retention; it is their interrelatedness which reveals the influence of management practices on workforce sustainability. The interrelatedness of these themes creates an overlap which is captured in Fig. 2.1.

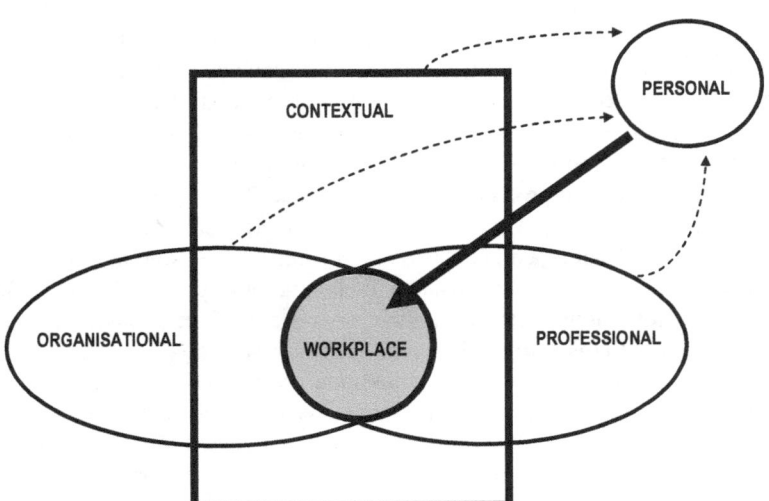

Fig. 2.1 Interrelatedness of themes describing key rewards and challenges for health professionals working in remote regions. The framework developed from this early conceptual version of the interrelated themes is published in Onnis and Pryce (2016, p. 47)

Figure 2.1 shows the way in which the four themes that emerged from the synthesis of the literature interact. The overlapping section in the centre is the 'workplace' (grey area). It is the place where all of the themes overlap and it is central to the argument that managers are the key to improving workforce sustainability. Why? Because the workplace is the place where everything comes together, it is the place where many decisions about whether to stay or leave are formed, and it is the place where the manager has the most influence. The two ovals that connect to give the workplace its shape represent the organisation and the profession. For health professionals, these themes represent two separate, yet connected sets of obligations and expectations, and create the boundaries for their work role. That is, they are restricted by the scope of their clinical practice (professional) and they are restricted by the policies put in place by their employer (organisational). That is, their profession may have guidelines for a particular safe practice, yet their employer may have a policy that has conflicting restrictions (i.e. best interest of the patient (professional) versus the most cost efficient way (organisational)). In remote areas, this is often more difficult because of the limited access available to other health professionals, e.g. many remote communities do not have resident doctors so nurses must seek clinical assistance from doctors outside of the remote community who may not understand the environment in which the nurse is caring for the client. In Fig. 2.1, the remote environment is represented by the rectangle; in the example just mentioned the doctor is not only outside the rectangle in which the nurse works, the doctor may have never been inside the rectangle. Finally, the circle with the arrow represents the health professional entering the workplace from outside of the remote context. This depicts the most frequent situation where a health professional enters the workplace from a location external to the remote community; however, it is acknowledged that some health professionals (e.g. local residents) do enter the workplace from within the rectangle.

In practice, what we are seeing are health professionals from outside the remote context making a decision to work in the remote context. They bring with them their 'personal' aspects, and arrive in the remote 'context' to work. Once in the remote workplace, aspects of their 'professional' practice interact with the 'organisational' policies and protocols in the workplace. These four themes are similar to those identified in other workforce studies (Cameron et al. 2012; WHO 2010). However, the HRM approach enables us to consider the person and their personal characteristics as a whole; then consider their relationship with the other factors, that is, their professional attributes, their organisational relationships, and their interaction with the environment in which they are working.

Figure 2.1 reveals that for health professionals working in remote regions, they enter remote workplaces with their own individual set of characteristics (Personal), their professional experience (Professional), their own perception of the employer (Organisational) and the environment in which they will be working (Contextual). This all influences their individual employment experiences and ultimately how long they remain in remote workplaces. Personal characteristics are unique to each health professional so person-fit, which is discussed in more detail in chapter four, provides multiple scenarios for the ways in which individual health professionals can work effectively in remote regions.

▥ Reflection

Using Fig. 2.1 think about how you fit into the remote workplace and draw a circle to show where you see yourself. Then, take a moment to reflect on Fig. 2.1 and your team.

1. Think about where you put your circle, and think about where members of your team would put their circles (in other words—where do they fit?).
2. Think about how this affects your workplace and the challenges that you face.

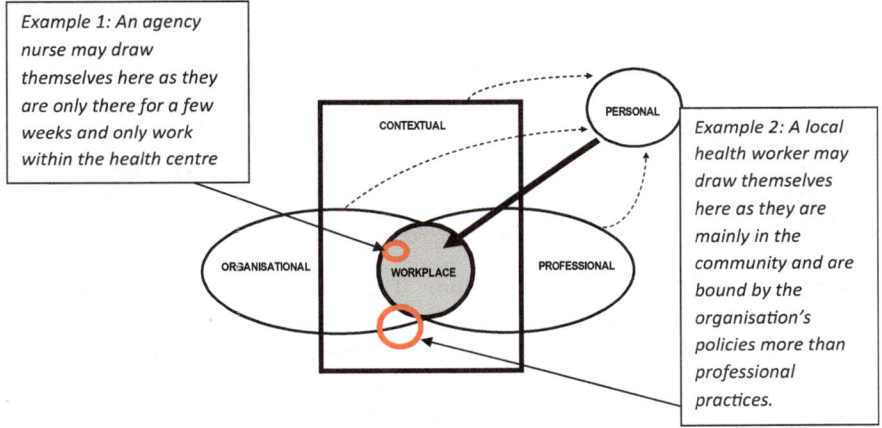

Take a moment to think about your past work experiences. List any conflicts you have experienced between your professional practice and the organisation's policies? If you are a manager, put an asterisk next to any that members of your team may experience as well.

1. How did you manage the conflict?
2. Would you use a different approach if faced with the professional/organisational conflict again?
3. If a member of you team came to you to discuss a professional/organisational conflict—what would you do? What advice would you offer them? What would you suggest they do?

▭ Manager's Toolkit

Once you have completed the reflection exercise, think about whether there are items that you can add to your Manager's Toolkit to help you to manage professional-organisational conflicts. Would it help to have documents, website addresses, or contact details in an easily accessible place? Could you write yourself a note about what worked for you as a reminder, if needed, in the future?

HRM and the Remote Workplace

In its broadest terms, HRM describes the development of systems within an organisational context so that the effective management of people is supported (Beer et al. 1984). More recently contemporary challenges such as globalisation and socialisation of work; the increase of knowledge workers; flexible work practices; and the rise in focus on the service industry, highlight that HRM theories and practices must be considered relative to time and context (Paauwe 2009). Six HRM concepts that will influence workforce sustainability for remote managers to consider are:

- **The mobilities paradigm**: where improvements in technology, changes in society and employment patterns support the development of a more globally mobile workforce.
- **Organisational commitment**: The commitment an employee has to the organisation for which they work which is often described as loyalty.
- **Occupational commitment**: The commitment that a person has to their profession. Occupational commitment is more personal and is often associated with a substantial personal investment in the profession (e.g. time and financial commitment to complete a university degree, time and expense to maintain and update skills).
- **Organisational citizenship behaviour**: These are behaviours that employees demonstrate that go above and beyond what would be expected of the average employee. Employees demonstrate organisational citizenship behaviours when they do not expect acknowledgement or compensation for their actions, e.g. staying late to assist a customer, picking a colleague up from the airport in their own personal car after hours.
- A **boundaryless career** describes a career which is not tied to one employer. For example, nurses can work for any employer as nursing skills are transferable to any employer. An increase in temporary employment arrangements, reduced job security and worker mobility has resulted in the rise of boundaryless careers.
- A **protean career** is often described as the reinvention of a career. It is where individuals take the skills and experiences gained while working in one career and use them to develop another career. The contemporary nature of work is making this type of career transition more common.

In recent times our traditional understanding of the 'workplace' has been tested. Technological advances, industrial and employment relations reform, Award modernisation, social, economical, environmental, political and cultural influences on how and when we work have been changing rapidly. Not since the industrial revolution has there been such rapid and continuous change in the work-related aspects of human life. It is in this context that the challenges for remote workforce sustainability are considered, as contemporary HRM challenges are just as relevant in geographically remote regions as they are in metropolitan cities. The increase mobility of workers has meant that many aspects of work which were traditionally associated with specific countries, are now globally relevant (Lippel et al. 2017; Newhook et al. 2011). As virtual workplaces, Telehealth, and international work-forces increase, the HRM challenges associated with managing such workforces will need to evolve at a similar pace. Some of the challenges rapidly evolving for contemporary employment have been experienced by remote health services earlier than their city counterparts (e.g. visa programs prescribing minimum periods for overseas trained clinicians to work in regional, rural or remote health centres prior to gaining work permits for city hospitals).

These workforce changes are reflected in contemporary management theories such as the new mobilities paradigm. This improved mobility is indicative of the factors that will continue to impact workforce sustainability in general, and may have a more substantial impact on remote regions already experiencing high turnover (Sheller and Urry 2004). Migration patterns that result in movement to urban areas are already common and for remote regions this mobility requires workforce planning and management (Kabene et al. 2006; Lippel et al. 2017; McGrail and Humphreys 2015; Newhook et al. 2011). At the same time, the remote workforce includes many itinerant workers who choose a travelling lifestyle, pre-ferring adventure, unique experiences or opportunities not available in the more competitive urban labour market (Nadkarni and Stening 1989). Further, increased temporary employment, and the dependence on agency staff in remote regions further complicate commitment and workplace relationships (Hudson and Inkson 2006). The changing structure of contemporary workforces and the mobility of individual employees create an opportunity for managers to improve retention through creative solutions, particularly creating a workplace that is attractive to health professionals. In a competitive labour market, these management practices may be the competitive edge needed to improve retention, particularly in remote regions where demand exceeds supply.

HRM evidence-based theories which include organisational commitment, occupational commitment, and organisational citizenship behaviours contribute to our understanding of high turnover in remote regions. These commitment beha-viours, or absence of commitment behaviours, can provide insight into employee motivation and can assist managers in knowing how to improve retention. Simi-larly, the rise in the boundaryless and protean career, where loyalty is reduced and personal success comes from self-directed career management, reinforces the propensity for occupational over organisational commitment (Hudson and Inkson 2006). Where health professionals are demonstrating commitment to their profes-sion, not their employer, there is an opportunity for managers to influence retention through creating a more satisfying place to work.

⊞ Reflection

If you had to choose between your profession or your employer where would your commitment lie? On the line below, mark where you think that you commitment lies. Is it closer to your employer or your occupation/profession?

Organisation   Occupation

Reflect on where you placed the mark. Was it an easy choice or was it difficult? Why? The answers to these questions will help you to understand how you make decisions about your employment. It will also help you, as a manager, to understand some of the decisions that your team make. For example, if a team member likes working with your team in the remote community, but cannot see any career advancement opportunities their occupational commitment may drive their decision to leave (whether they know it or not). In contrast, some team members may put up with poor employment conditions if it means that they can do the work (occupation) that they love.

As a manager, you may or may not be able to influence turnover and retention; however, if you reflect of the type of commitment that a team member is demonstrating you may be able to improve retention through opportunities that reinforce the type of commitment they demonstrate. For example, someone who demonstrates organisational commitment may value organisational recognition of their good work through an award or by publicly acknowledging them.

Manager's Toolkit

In the past month, have you or any members of your team demonstrated organisational citizenship behaviour? Is your team just meeting the minimum requirements needed or do they go the extra mile? As a manager how do you acknowledge organisational citizenship behaviour?

For the next week, make a list of any organisational citizenship behaviour you observe and how this is acknowledged. Make a note of any changes that you see when this behaviour is acknowledged. What worked? What didn't work? Remember, some employees are not looking for recognition, they are usually internally motivated, so 'thank-you' may be enough.

You may like to add a list of the ways you acknowledged organisational citizenship behaviour that received a positive response. Over time you can add to this list until you have a personalised list of ways that you like to acknowledge your team when they display organisational citizenship behaviour in the workplace. Sometimes in a busy workplace, the small things are overlooked, so keeping this list in your Manager's Toolkit will serve as a prompt that can be used when needed.

Decision Making in the Remote Workplace

> 'The decision to locate in a rural practice setting occurs largely from outside that setting. The decision to remain takes place from within the practice setting and arises from the stream of experience there.'
>
> (Cutchin 1997, p. 1662)

In this and future chapters, creative solutions are explored as well as the management theories and frameworks that can guide and inform remote managers in navigating the contemporary HRM challenges in modern workplaces. These include, but are not limited to, recruitment practices, personal and professional support and people management; all of which will positively impact workplace culture, morale and operational functioning (Rosete 2006). In understanding the significant influence of 'the workplace' on workforce sustainability in remote regions, it is imperative to consider that for many remote health professionals, the 'remote workplace' is the majority, if not their entire understanding of the organisation for whom they work. Therefore, it is from the remote workplace and with the people that they work, that they will make their decision about when to leave the organisation.

Through effective management practices, managers can create a workplace culture conducive to attracting, recruiting, developing and supporting the most appropriate workforce for that particular remote workplace. If the past reflects the future, this is not about a set of fixed HRM policies and procedures, it will necessitate, that the remote manager understands the remote context and can interpret HRM policies so that they are effective in the context in which they are implemented. This is of particular relevance in remote workplaces, where small groups of people who live and work together in relative isolation necessitates some degree of localisation that may not usually be considered in larger urban workplaces.

Managers Are the Key to Workforce Sustainability

For health professionals, working in remote regions is challenging; however, for managers there are additional challenges (Greenwood and Cheers 2002). Managers experience an additional burden undertaking the duties of planning, supporting and motivating others in a complex environment often with limited management experience, training or support (Greenwood and Cheers 2002; van der Heijden et al. 2009; Lenthall et al. 2009). If health professionals make decisions about whether or not to remain based on their experience in the 'remote workplace', it follows that the manager, the most influential person in that workplace, will have a significant impact on turnover (Cutchin 1997; Fisher and Fraser 2010). However, managers need support in improving their ability to manage remote health professionals, and this support needs to extend beyond the current management workforce. Attention should also be focused on supporting health professionals in remote regions who have the potential to be the next managers. These health professionals understand the remote context; they are making decisions from within the work environment, so retention through ongoing professional support and succession planning is vital for organisations facing workforce shortages through high turnover. If as Buckingham and Coffman (1999) proclaimed turnover is a management issue; it follows that managers are the key to workforce sustainability.

References

Alasia A, Bédard F, Bélanger J, Guimond E, Penney C (2017) Measuring remoteness and accessibility—a set of indices for Canadian communities, Catalogue no. 18-001-X

Battye KM, McTaggart K (2003) Development of a model for sustainable delivery of outreach allied health services to remote north-west Queensland, Australia. Rural and Remote Health 3:194

Beer M, Spector B, Lawrence PR, Mills DQ, Walton RE (1984) Managing human assets. The Free Press, New York

Belaid L, Dagenais C, Moha M, Ridde V (2017) Understanding the factors affecting the attraction and retention of health professionals in rural and remote areas: a mixed-method study in Niger. Hum Resour Health 15(1):60

Bent A (1999) Allied health in Central Australia: challenges and rewards in remote area practice. Aust J Physiother 45(3):203–212

Birks M, Mills J, Francis K, Coyle M, Davis J, Jones J (2010) Models of health service delivery in remote or isolated areas of Queensland: a multiple case study. Aust J Adv Nurs 28(1):25–34

Buckingham M, Coffman C (1999) First, break all the rules. What the worlds greatest managers do differently. Simon and Schuster, London, UK

Cameron P, Este C, Worthington C (2012) Professional, personal and community: 3 domains of physician retention in rural communities. Can J Rural Med 17(2):47–55

Cutchin MP (1997) Community and self: concepts for rural physician integration and retention. Soc Sci Med 44(11):1661–1674

Devine S (2006) Perceptions of occupational therapists practising in rural Australia: a graduate perspective. Aust Occup Ther J 53(3):205–209

Fisher KR, Fraser JD (2010) Rural health career pathways: research themes in recruitment and retention. Aust Health Rev 34(3):292–296

Gardiner M, Sexton R, Durbridge M, Garrard K (2005) The role of psychological well-being in retaining rural general practitioners. Aust J Rural Health 13(3):149–155

Greenwood G, Cheers B (2002) Doctors and nurses in outback Australia: living with bush initiatives. Rural Remote Health 2:98

Hays R, Wynd S, Veitch C, Crossland L (2003) Getting the balance right? GPs who chose to stay in rural practice. Aust J Rural Health 11(4):193–198

Hegney D, McCarthy A, Rogers-Clark C, Gorman D (2002a) Retaining rural and remote area nurses. The Queensland, Australia experience. J Nurs Adm 32(3):128–135

Hegney D, McCarthy A, Rogers-Clark C, Gorman D (2002b) Why nurses are attracted to rural and remote practice. Aust J Rural Health 10(3):178–186

Hegney D, McCarthy A, Rogers-Clark C, Gorman D (2002c) Why nurses are Resigning from rural and remote Queensland health facilities. Collegian 9(2):33–39

Hudson S, Inkson K (2006) Volunteer overseas development workers: the hero's adventure and personal transformation. Career Dev Int 11(4):304–320

Humphreys JS, Jones MP, Jones JA, Mara PR (2002) Workforce retention in rural and remote Australia: determining the factors that influence length of practice. Med J Aust 176(10):472–476

Jackson D, Clare J, Mannix J (2002) Who would want to be a nurse? Violence in the workplace—a factor in recruitment and retention. J Nurs Manag 10(1):13–20

Kabene SM, Orchard C, Howard JM, Soriano MA, Leduc R (2006) The importance of human resources management in health care: a global context. Hum Resour Health 4(20)

Kelly K (2000) Improving quality of life of remote area nurses. Collegian 7(4)

Kent-Wilkinson A, Starr L, Dumanski S, Fleck J, LeFebvre A, Child A (2010) International nursing student exchange: rural and remote clinical experiences in Australia. J Agromedicine 15(1):58–65

Knights JA, Kennedy BJ (2005) Psychological contract violation: impacts on job satisfaction and organizational commitment among Australian senior public servants. Appl HRM Res 10 (2):57–72

Kruger E, Tennant M (2005) Oral health workforce in rural and remote Western Australia: practice perceptions. Aust J Rural Health 13:321–326

Lenthall S, Wakerman J, Opie T, Dollard M, Dunn S, Knight S, MacLeod M, Watson C (2009) What stresses remote area nurses? Current knowledge and future action. Aust J Rural Health 17 (4):208–213

Lenthall S, Wakerman J, Opie T, Dunn S, MacLeod M, Dollard M, Rickard G, Knight S (2011) Nursing workforce in very remote Australia, characteristics and key issues. Aust J Rural Health 19(1):32–37

Lippel K, Johnstone R, Baril-Gingras G (2017) Regulation, change and the work environment. Relations Industrielles/Ind Relat 72(1):3–16

McGrail MR, Humphreys JS (2015) Geographical mobility of general practitioners in rural Australia. Med J Aust 203(2):92–96

McGrail MR, Humphreys JS, Joyce C, Scott A, Kalb G (2011) Rural amenity and medical workforce shortage: is there a relationship? Geogr Res 49(2):192–202

Nadkarni S, Stening BW (1989) Human Resource management in remote communities. Asia Pac J Hum Resour 27(3):41–63

Newhook J, Neis B, Jackson L, Roseman S, Romanow P, Vincent C (2011) Employment-related mobility and the health of workers, families, and communities: the Canadian context. Labour (Spring):121–156

O'Toole K, Schoo AM (2010) Retention policies for allied health professionals in rural areas: a survey of private practitioners. Rural Remote Health 10:1331

Onnis L, Pryce J (2016) Health professionals working in remote Australia: a review of the literature. Asia Pac J Hum Resour 54:32–56

Opie T, Lenthall S, Wakerman J, Dollard M, MacLeod M, Knight S, Rickard G, Dunn Sandra (2011) Occupational stress in the Australian nursing workforce: a comparison between hospital based nurses and nurses working in very remote communities. Aust J Adv Nurs 28(4):36–43

Paauwe J (2009) HRM and performance: achievements, methodological issues and prospects. J Manage Stud 46(1):129–142

Rosete D (2006) The impact of organisational values and performance management congruency on satisfaction and commitment. Asia Pac J Hum Resour 44(1):7–24

Santhanam R, Hunter E, Wilkinson Y, Whiteford H, McEwan A (2006) Care, community, capacity: rethinking mental health services in remote indigenous settings. Aust J Prim Health 12(2):51–56

Sheller M, Urry J (2004) The mobilities paradigm. Environ Plan A 38:207–226

van der Heijden BIJM, van Dam K, Hasselhorn HM (2009) Intention to leave nursing: the importance of interpersonal work context, work-home interference, and job satisfaction beyond the effect of occupational commitment. Career Dev Int 14(7):616–635

Wakerman J, Humphreys JS, Wells R, Kuipers P, Jones JA, Entwistle P, Kinsman L (2009) Features of effective primary health care models in rural and remote Australia: a case-study analysis. Med J Aust 191(2):88–91

World Health Organisation (WHO) (2010) Increasing access to health workers in remote and rural areas through improved retention. WHO Press, France. http://www.searo.who.int/nepal/mediacentre/2010_increasing_access_to_health_workers_in_remote_and_rural_areas.pdf. Accessed 5 Sept 2017

Sustainable Remote Health Workforces

3

Mother Teresa's philosophy was relatively simple, yet powerful. She made people feel valued by genuinely paying attention to them.

Dr. Tim Baker, Managing Director, Winners at Work.
(Baker 2014, p. 136)

Key Messages

- A sustainable remote health workforce is about an appropriate mix of health professionals with suitable personal characteristics and professional attributes to meet the remote populations' needs.
- Health professionals working in remote regions propose that future health workforce sustainability is achievable with effective management practices focused on people, practice and place.
- Irrespective of geographical location, it seems that infrastructure, resources, and community engagement together with the health provider's systems influence the effectiveness of management practices. Hence, management practices influence the outcomes that can improve workforce sustainability.

What Is Workforce Sustainability?

This chapter focuses on workforce sustainably, a theme that is woven through each chapter of this book. Workforce sustainability is an ambiguous term that is open for interpretation. In fact, Buchan et al. (2011) proposed that making a policy statement on health workforce sustainability is a lot easier than agreeing on what it actually

© Springer Nature Singapore Pte Ltd. 2019
L. Onnis, *HRM and Remote Health Workforce Sustainability*,
Management for Professionals, https://doi.org/10.1007/978-981-13-2059-0_3

means. As a consequence, there is no agreed definition of what health workforce sustainability means in practice, and there are differing views on whether it is achievable under any working definition (Buchan et al. 2011). Therefore, it is important to define what is meant in this book by workforce sustainability in the context of remote health workforces.

This book adopts the definition of Kossek et al. (2014, p. 299) who define a sustainable workforce as, 'one whose employees have the positive energy, capabilities, vitality, and resources to meet current and future organizational performance demands while sustaining their economic and mental health on and off the job.' In other words, members of the workforce have the ability to meet the current and future needs of the organisation.

Humphreys et al. (2006, p. 33) say that in 'the rural and remote health context, the concept of sustainability refers to the ability of a health service to provide ongoing access to appropriate quality care in a cost-efficient and health effective manner'. Therefore, in terms of a remote health workforce, sustainability refers to the continual supply of competent health professionals, all of whom provide health services through practices suitable for the remote context in which they work, as well as aligning their personal values with those of the organisation, and the communities in which they live and work (Onnis 2016). Ultimately, sustainability is built on a strong foundation, where managers and leaders meet the challenges and respond to the opportunities, ensuring that the needs of all involved continue to be fulfilled. Hence, for sustainable remote health services, remote managers must take 'account of the social, economic and environmental dimensions influencing sustainability' focusing on the quality of care, access to health services, and the cost of providing health services (Humphreys et al. 2006; Onnis 2016).

In remote regions worldwide, efforts to establish sustainable health workforces have been hindered by many factors that complicate what is already a complex environment. An environment shaped by health systems, government priorities, economic constraints fuelled by competing demands for reduced funding, stakeholder interests, changing community and individual health needs and for many organisations, a reduced pool of competent and capable health professionals willing to work in remote regions. As a result, attracting and retaining health professionals in remote regions is a global challenge; one that is made worse by the disparity of distribution of health professionals between urban and remote areas (Wakerman and Humphreys 2012; WHO 2010). Workforce instability, particularly workforce shortages intensify the existing challenges of workforce attraction and retention faced by remote managers.

In the past, researchers have examined turnover and retention of particular occupation-based remote health workforces, including nurses, doctors and allied health professionals (Garnett et al. 2008; Hays et al. 2003; Hegney et al. 2002a; Humphreys et al. 2002, 2006; Kruger and Tennant 2005; O'Toole and Schoo 2010). This chapter examines the challenges of workforce sustainability from the perspectives of health professionals of various professions who are currently working in remote regions across (allied health, Indigenous health workers, medical, and nursing), an area where gaps have been identified in our knowledge about retention and workforce sustainability (Buykx et al. 2010; Onnis and Pryce 2016; WHO 2010). This approach seeks to understand aspects of workforce sustainability that extend beyond the boundaries of a particular profession. The characteristics of a sustainable remote health workforce emerged through applying a HRM perspective to examine how the current remote health professionals describe sustainable remote health workforces.

Perspectives from the Current Remote Workforce

'A sustainable workforce is one which is not person-dependent but at the same time values the individual skills, experiences and ideas a person can bring to a role… It is one that is able to provide continuous, reliable and safe care to patients… because staff are appropriately skilled, oriented, supported and rewarded. It is one where staff movement is pre-empted, planned and refilled in a timely and appropriate manner. It is one that doesn't rely on agency and locum staff, but grows a local workforce wherever possible and provides the same incentive packages as those afforded outsiders. It is one which is open to innovation and challenges the status quo with regards to new models of care and expanded scopes of practice, and recognises the skills all individuals bring to the cause. It can be achieved by employing people who are passionate about their job and love a rural/remote lifestyle' Remote Health Professional.' (Onnis 2016, p. 1)

Overall, the characteristics of a sustainable remote health workforce, as described by health professionals working in remote regions at the time of the study, were diverse and varied; however, there were many common characteristics. The study revealed three themes—people, practice and place. The remainder of this chapter will focus on what managers can do to improve remote health workforce sustainability arranged under the themes: people, practice and place. Please see Appendix A for more information about the research methodology.

ⓘ Reflection
Take a moment to reflect on your current workplace. How would you describe workforce sustainability for your current workplace?

Create a table and list what is needed to improve workforce sustainability in your workplace in the first column. In the second column make a note of what you could do to improve workforce sustainability. Be as specific as possible by writing clear actions (examples are provided in the following table).

What is needed to improve workforce sustainability	What I can do improve workforce sustainability
Example Reduce the gap between staff leaving and new staff arriving	*Example* (1) Start the recruitment process sooner (2) Create a list of Agency staff that would be suitable to fill vacancies

Research Findings About the 'People' Aspects of Workforce Sustainability

The theme 'people' comprises 'personal characteristics' and 'professional attributes' of individual health professionals that can improve workforce sustainability (Table 3.1). Similar aspects of personal characteristics and professional attributes for individual health professionals were described in the study; however, there were differences in terms of what was considered to be the priority, with no consensus as

Table 3.1 Management practices for the 'People' theme that improve workforce sustainability
(Some of the ideas contained in this table were previously published in Onnis 2016)

Personal characteristics	Professional attributes
To improve *person-fit*, managers can: • Prepare new recruits for the reality of remote practice • Assist employees to meet their needs and aspirations • Be more aware of the challenges of living in remote areas • Employ health professionals that like living and working in remote areas	To improve *competence*, managers can: • Create teams with a mix of experience and qualifications • Recruit health professionals who work well in multi-disciplinary clinical teams • Value knowledge and experience gained over time • Fill management positions with people who have previous remote health experience • Demonstrate leadership in management
To improve *individual self-care*, managers can: • Assist health professionals (and themselves) to recognise the early warning signs of fatigue and stress • Recruit for resilience and people whose energy and passion for remote work is sustainable • Promote a reasonable work-life balance • Ensure that health professionals (and themselves) take regular breaks from the remote area • Recruit people who gain true joy from their work	To improve *individual professional development*, managers can: • Facilitate access to regular professional development for all health professionals • Encourage people to participate in formal and/or informal mentoring programs • Facilitate access to customised remote professional development • Facilitate exchanges/rotations with health professionals in major centres • Support 'grow your own workforce' strategies
To improve *relationships*, managers can: • Establish community relationships • Work collaboratively with community without prejudice • Recognise and reinforce the benefits, for self and clients, from engaging with the local community	To improve *career paths*, managers can: • Support career development • Recognise the large (and broad) skill set of remote health professionals • Encourage health professionals on fixed term contracts to remain • Ensure that short-term workers are contributing to workforce sustainability not creating instability • Recruit health professionals who have participated in rural and remote graduate training programs

to whether it is person-fit, competence, or relationships. Of note, most health professionals suggest that 'person-fit' is essential. They suggested that employing the right people for remote regions was vital; however, there was no consensus about the characteristics of the right people, or how to recruit them. While this is disappointing for those who were looking for a checklist of the characteristics suitable for recruiting the 'right people' to work in remote regions; it is not surprising given the diversity of human beings, health services and remote communities. In other words, the right fit for one health service in one community, may not be the same for a different health service in a different community.

Table 3.1 contains suggestions about actions that remote managers can take to improve workforce sustainability. For example, to improve 'person-fit' managers can prepare new recruits for the reality of remote practice. To localise the implementation of this suggestion, the remote manager must consider what they are currently doing to prepare new recruits and then consider how this could be improved. The manager may do this alone, in conjunction with the team or they may devise a way to collect this information from new recruits after they commence. Putting a continuous quality improvement (CQI) process in place like this, can improve workforce sustainability and will help the manager to understand the personal characteristics of each new team to further understand the personal characteristics that are associated with the health professionals who are a better fit for their health service and community. Working through Table 3.1 remote managers can consider each of the suggested management practices, how they could benefit workforce sustainability in their workplace and what actions they can take to implement the changes. For each item, there is an opportunity to reflect on applicability of the suggested management practice to their current situation.

Ⓘ Reflection
Reflect on the information in Table 3.1 and note any relevant to your workplace.

1. Which aspects improved workforce sustainability in your workplace?
2. Which are giving you insights into personal characteristics that could improve the 'people' aspects of the recruitment process.

Manager's Toolkit
Using the first example discussed from Table 3.1, managers can develop a form or a short questionnaire to ask health professionals on commencing how they (the remote manager) could have better prepared them for commencing work in the remote community. Then, the remote manager can collate this information and use it to better prepare the next new team member. This form/checklist can be added to your Manager's Toolkit and is transferable, that is, it can be used in any remote community.

Further ideas for your Manager's Toolkit

- Prepare new recruits for the reality of remote practice—find out whether your organisation already has information; develop an information sheet/booklet; adapt resources from other organisations. For example, Northern Territory's Remote Health's booklet 'Remote Ready' available at http://www.remoterecruitment.nt. gov.au/docs/remote_ready.pdf and Queensland Health's Employment, rural and remote guide at https://www.health.qld.gov.au/employment/rural-remote/.

- Be aware of the challenges of living remote—develop strategies to support team members to work through the challenges of working in remote communities.
- Consider including questions in your recruitment process to help you to identify health professionals that like living and working in remote areas. Take a moment to write a few suitable questions and put them in your Manager's Toolkit.
- Reflect on the best mix of experience, characteristics and qualifications for your team. When recruiting map out the experience, characteristics and qualifications of your current team identifying any skill gaps. Then determine how to best recruit the mix of skills, characteristics and experience your team needs.
- Self-care—develop a system to ensure that your team members (and you) take regular breaks from the remote area.
- Add resources to your Manager's Toolkit to help you (and for you to assist your team) to recognise the early warning signs of fatigue and stress, and to monitor your mental health and wellbeing. Some potential resources are listed.

🪧 Resources

There are various resources available, listed here are a few to get you started.

Moodprism—An app to map your mood (Beyond Blue)
http://www.moodprismapp.com/

Australian Unity Wellbeing Index
https://australianunity.asia.qualtrics.com/jfe/form/SV_3Wrr9BX1iBvSJfL

My Compass (The Black Dog Institute)
https://www.mycompass.org.au/

Work-life balance resources
https://www.healthdirect.gov.au/work-life-balance

IN PRACTICE BOX 3.1: Changing the Nursing Profile from Agency Nurses to Improve Workforce Sustainability

Sandra's Story
We went from depending on agency nurses to fill the gaps to a stable workforce where health professionals and their families integrated into the remote community.

When I started as the manager of the remote hospital we survived on agency staff. Turnover was high and every three months we would have a new group of nurses coming through. We had 15 positions but we never had more than five permanent staff at one time. It doesn't sound like many but when you have key people moving through the health system it wears down the permanent staff and it wears down the community as well. They see a constant parade of nurses passing through the hospital.

The Remote Workplace

Prior to coming here I had been working in the emergency department at a city hospital for quite a while so I felt quite skilled to work in isolated areas. However, when I arrived in the remote town, the hospital was running on four nurses and some agency nurses and very soon after I arrived, two of those permanent nurses resigned. So I was left with a base workforce of two permanent nurses. When I looked at the living conditions for nurses, they were very basic and people were not entirely happy with them. At that time staffing was very problematic. Although the team worked really well together, it can be quite tiring for staff when they are recalled to the hospital regularly. They were often recalled if there was a psychiatric patient, or a cardiac patient that needed to be flown out. Nurses still have to do the one-on-one care until you can get them out of the remote town, so not having enough relief staff was one of the main staffing issues. In remote towns, you don't have the flexibility that you have in the metropolitan areas; where you can just get an agency nurse in for half a day when you are short of staff.

I found that often there was a big difference in what agency nurses expected that they would be doing in a remote hospital and what we expected them to do. Consequently, there was also a difference in their skill level compared to the skill level we expected. So we were very keen to make sure that we attracted nurses that had the appropriate skill level for a remote hospital, because sometimes they would arrive and did not have the skills that we needed, so that was disappointing but most times we could work around it. We would need to work them in areas where they did have the skills and then make sure that they did not come back. This happened often enough for it to be a challenge. The other challenge was that sometimes a nurse's behaviour after work hours can be really problematic in small remote communities. We had to send a few nurses off quick smart because they just did not have the right behaviour out of work. In remote towns, you cannot go to the pub, get drunk and dance on tables and then come to work the next day and pretend that nothing happened. It just does not work in small communities; it is bad for the hospital, and it is bad for the other nurses that work in the hospital too.

The Recruitment Strategy

Over time, I realised that we needed to attract families to this area for the benefit of the health service and the people in our community. The community just see the nurses as people who are single, people coming and going, people who come here for a good time and are not really here to improve the health of the community. I wanted to change that perception and so I started by looking at the accommodation, which was all set up for single people, located near the hospital and meant that nobody would stay in town. Over a period of about 12–18 months, I started to change the profile of the nursing staff in the hospital by recruiting overseas trained nurses, which was really a bit of a gamble because we are in an isolated area. I started to recruit people from overseas and they came with a contract for a fixed period of time. Initially, I recruited three nurses from India, based on a feeling that if I recruited one they might feel a bit isolated, so I recruited three. They have subsequently married, two of them have children and they are still living in that remote town, and that was three years ago.

Workforce Stability

So that was the start of the transition to change the nursing profile of the remote hospital and it was quite successful. We went on to recruit two more nurses from India after about eight months and then another four after another twelve months. They are all still there, except one of them who left to move interstate with her husband. This strategy gave us a stable workforce and then we were able to build the health service and integrate quality improvements into the service we were providing. It also meant that the people in the remote town started to get to know the nurses. Previously, people were churning through the hospital at such a rate the community probably felt that we did not care. So, there was a big change at the hospital which enabled us to stabilise the staffing and then the community started to accept the hospital.

So, I think that it comes down to what we value. If we are living in the community, we need to be part of the community and to be part of the community we need to demonstrate that we have the same values as the community; we also have a family, so we send our children to the same school, we use the same shop and we all have access to the same community resources.

Localising HRM Policies

The nurses were recruited through the overseas training program registered with ATRA. We brought them straight to the remote town and we orientated them ourselves. We trained them in the skills needed to work in the remote hospital which was crucial in the success of the appointments, because it was about making sure that the people we recruited have the skillset to work competently in the remote hospital but also have the capacity to develop to meet future demands. It was very clear that I was getting nurses with potential. They may not have had all the skills that I needed them to have initially, but I made sure that they were upskilled quickly. This is where my senior management shone. They were able to allow me a bit of latitude to get these nurses skilled up, knowing that the investment would give us skilled nurses for a reasonable period of time. Furthermore, the overseas trained nurses provided the workforce stability we needed to improve the quality of our health services.

So, from my point of view it was successful.

Research Findings About 'Practice' Aspects of Workforce Sustainability

The theme 'practice' comprises two aspects of professional practice that health professionals believe contribute to workforce sustainability: clinical practices and workplace practices. Clinical practices includes aspects of health service delivery, and workplace practices described aspects of organisational and health system policies and practices. All of which according to remote health professionals contribute to the sustainability of remote health workforces (Onnis 2016). Health professionals found that turnover often led to recurrent vacancies which impacted on their capacity to provide quality health services, with one health professional saying, 'you need to be realistic with sustainable, but I guess it would be something like ensuring that 95% of your positions remained filled' (Onnis 2016). Health professionals suggest that managers could support workforce sustainability through clearer communication, genuine understanding of the challenges of remote work environments including adequate health and safety, and by maintaining reasonable levels of core staff, through backfill and expediently filling vacancies (Bent 1999; Devine 2006; Fisher and Fraser 2010; Kruger and Tennant 2005).

There were varied perspectives about whether continuity of care was dependent on models of health service delivery, with some people proposing that continuity of care was closely associated to continuity of health professionals. As one health professional explained, 'our poor clients, they have such a change of faces, and it takes time to develop the relationship' that supports continuity of care. Others suggested that the way forward would be the fly-in, fly-out (FIFO) model, saying that, 'A sustainable workforce does not mean people who work in one position/one site for a long time. It means the positions give individuals the opportunity to grow in their field'. These varied perspectives suggest localised approaches to working sustainably are needed to maintain continuity of care. Management practices contribute to the sustainability of the remote health workforce, particularly in the implementation of policies for filling vacancies, offering backfill, attracting health professionals, remuneration and financial incentives, employment patterns, and models of practice (Table 3.2).

For remote managers, high turnover and workforce instability is both time and energy consuming. Firstly, there is emotional strain brought about by the continual ending and commencing of working relationships. Also, there is the additional work of covering vacancies, recruitment processes and orientation for new health professionals. As such, many remote managers are seeking opportunities to improve workforce sustainability so that their energy can be focused on operational matters, such as quality health services and meeting community health needs.

Table 3.2 **Management practices for the 'Practice' theme that improve workforce sustainability** (Some of the ideas contained in this table were previously published in Onnis 2016)

Clinical practices	Workplace practices
To improve the *model of practice*, managers can: • Ensure that the FIFO workforce supports the remote-based workforce • Ensure remote workforces are not dependent on FIFO and/or agency staff • Use FIFO as a viable workforce solution when attracting resident staff is not possible • Aim for a consistent workforce	To improve *work systems*, managers can: • Provide and encourage regular leave • Ensure that city-based managers understand the unique geographical differences • Ensure there are sufficient core permanent staff to balance FIFO and agency staff • Be responsive, personalise support and ensure there is no workplace bullying • Ensure open communication and employee voice • Ensure health professionals feel valued and respected • Ensure adequate administrative support • Reduce inefficiency to reduce time-costs • Base promotions on skills not longevity • Create prepared, empowered, autonomous and well supported workforces
To improve the *continuity of staff*, managers can: • Encourage and support health professionals to stay in remote regions • Strive to lower turnover so that people who know the community and the system remain longer • Develop succession planning strategies to improve sustainability	To improve *reward systems*, managers can: • Ensure the financial rewards are suitable • Offer incentives that improve retention • Ensure remuneration and compensation is equitable and fair for the entire workforce • Offer incentives for long-term employees not just new ones • Offer incentives to employees from the local area, as well as from outside the area

ⓘ Reflection

It may be worth considering examining aspects of job design in your workplace.

1. How do you determine who does which tasks in your workplace?
2. Do your fly-in fly-out/drive-in drive out workers support your remote-based workers? Or vice versa? Are there any changes that you could make that would improve the sustainability of your workforce (i.e. reduce pressure, changing the way you allocate tasks)?

It may be worth mapping work activities to team members, and also identifying the 'additional remote activities', that is, the work activities that need to be completed because of your remote location; activities they would not be part of a clinician's regular work role in an urban health centre (e.g. checking the oil in the ambulance). Then think about how you would approach a discussion about these activities with your team. *Please note: Do not walk into this discussion unprepared.* It will be important to be able to gauge the work climate (so do not proceed if tensions are high) but if the climate is right, an open discussion with transparent decision-making about these type of work activities can help to improve morale, team cohesiveness and productivity. Being inclusive in decision-making is an opportunity to reduce feelings of exclusion within the team. However, it is not always appropriate to include the team in decision-making so consider this suggestion including possible repercussions (and possible responses you can make) to any negativity before implementing. Review the available resources about job design and put any that appeal to you in your Manager's Toolkit. Taking time to considering job design can be useful for managers who need to build or restructure a team, during the early stages of the recruitment process, when conducting performance reviews and career planning.

Write in your reflective journal so that you can note what went well and what could be improved so that next time you can be better prepared and will feel more confident about when and how to use job design to improve workforce sustainability.

☂ Resources

There are four aspects of job design that managers can use to redesign, reconfigure or reshape a particular role.

- **Job simplification** is the process of reconfiguring positions so that the employee has a small number of tasks to perform with a narrow scope for each one. Managers are usually more directive and quite autocratic in managing positions that have been simplified.
- **Job rotation** is the practice of transitioning employees through a set of jobs in a planned sequence, usually with a defined period attached (e.g. 6 monthly, annually). This practice can be a good way of skilling a small workforce across several roles, it can expose employees to other tasks, and can be used for succession planning.
- **Job enlargement** is used to describe the redesign of a position so that it covers a wider variety of similar tasks. Job enlargement can be used to keep employees motivated and interested by making a position more challenging (i.e. an increased variety of tasks).
- **Job enrichment** is the process of upgrading the task mix for the position in order to significantly improve the employee's potential for growth, achievement, responsibility, and recognition.

While job design is a useful management tool for employee engagement and the development of interesting positions that can support the growth and development of employees; any change of tasks much be considered within the scope of the current Industrial Agreement for your workplace. As such, it is recommended that remote managers discuss any proposed changes to a position with an HR or Employment Relations Advisor/Officer before implementing any changes that may have industrial implications (e.g. salaries, responsibilities).

⑪ **Reflection**

1. Take some time to reflect on the workforce model at your workplace.
2. Does it work?
3. Could it be improved?

Do you have the capacity and/or opportunity to make any changes to the current workforce model?

As a remote manager, you are often caught between the frontline workers and senior managers, which is often not an easy space to occupy. So be realistic about what you can do, sometimes small changes make a big difference, and sometimes it is about being prepared when an opportunity presents so write them in your reflective journal (hopefully you are keeping one by now and if not it is never too late to start!)

Research Findings About 'Place' Aspects of Workforce Sustainability

The third theme that emerged in the study related to the aspects of workforce sustainability that were associated with a connection to place, and aspects of the physical work environment (e.g. infrastructure and resources) (Table 3.3). Extending on the suggestions about person-fit, some health professionals emphasised that community acceptance is important as well, saying,

> 'A sustainable remote health workforce is actually a workforce that is developed from the community and it is a workforce that the community accepts as well. I think there's not enough attention paid to the right fit in a community.' (Onnis 2016)

Health professionals suggested that a sustainable remote health workforce must be 'a workforce that the community accepts'; and that it 'needs to have people that are living in the community that belong to the community' (Onnis 2016). In addition, workforce sustainability is supported through greater connectivity between the different service providers who work within the same communities. The most frequently raised infrastructure concern was accommodation in remote regions with health professionals suggesting that addressing accommodation inadequacies was essential for workforce sustainability.

⑪ **Reflection**
Accommodation and incentives are difficult employment aspects for remote managers to manage for several reasons which are discussed in more detail in chapter seven. However, this is an opportunity to reflect on accommodation and incentives in terms of connection to place. Think about the policies and practices in your workplace.

Table 3.3 Management practices for the 'Place' theme that improve workforce sustainability (Some of the ideas contained in this table were previously published in Onnis 2016)

Connection with place	Infrastructure
To improve *community connections*, managers can: • Ensure that the workforce is integrated into the community • Recruit a workforce that call rural and remote Australia home, and not just an adventure • Ensure that there is mutual respect for the community, culture and beliefs • Employ local employees who have firm local connections (family and friends) • Create a balance between the needs of the health service and community	To improve *connectivity*, managers can: • Take a lead role in ensuring that health organisations work together towards workforce sustainability • Communicate with managers at other remote health services to reduce duplication of services and therefore reduce the number of vacant positions • Work with their peers to create a recruitment pool across the remote regions
To develop a *local workforce*, managers can: • Put local community members at the forefront of healthcare • Recruit locally or from 'like remote areas' • Recruit and develop local people • Value the contribution of a local workforce • Collaborate with other community and health services to improve workforce sustainability • Strive to develop a workforce that does not require recruitment from outside the area	To improve *resourcing*, managers can: • Provide housing for local employees as well as employees coming from other areas • Provide safe, affordable housing in a quiet part of the community • Offer accommodation incentives for permanent staff with their own homes • Offer consistent access to free accommodation • Offer accommodation suitable for singles, couples and families • Ensure there is sufficient infrastructure to support health service provision • Provide modern facilities, up-to-date technology and access to expert knowledge • Provide technology and equipment comparable to that of urban and regional centres

1. How does the team member's connection to place influence their expectations about accommodation?
2. Are team members from the local areas and/or those residing in the remote area prior to commencing work with the health service compensated in the same way as those employed from outside the local area?
3. What are the differences and how do they influence workforce sustainability?
4. Are there any actions that you can take to minimise the negative influence and/or promote the positive influence that accommodation and incentives have on the local workforce?

Towards Sustainable Remote Health Workforces

In the study, health professionals described sustainable remote health workforces in terms of the people who have an interest in it being successful and the people who manage remote health workforces. While some health professionals suggested that it is necessary to improve access to resources and the quality of infrastructure such as housing and clinic equipment; others suggested areas where sustainability can be improved through more equitable distribution of resources (Buykx et al. 2010; Onnis 2016). In particular, they suggest that where there is inequity in financial incentives offered to health professionals from different professions (e.g. nursing and allied health) this is likely to influence retention, particularly where health professionals work in multi-disciplinary teams and make a comparable contribution towards health service provision (Hegney et al. 2002; Onnis 2016; Santhanam et al. 2006; WHO 2010). Similarly, health professionals 'highlight differences in incentives and benefits offered to attract new health professionals compared to those offered to those residing in the local community. This disparity and perceived inequity appears to make local health professionals feel less valued, and appears to not be rewarding those that are providing the desired workforce stability' (Onnis 2016, p. 8). Hence, when management practices suitable to the context are combined with appropriately targeted retention incentives and rewards, they are more likely to influence workforce sustainability in the long-term (Fisher and Fraser 2010; Onnis and Dyer 2017).

A solutions-focused approach reveals possibilities for policies, and management practices that will have a positive influence on the sustainability of remote health workforces. Remote regions are geographically isolated; however, with technological advances, improvements to regional infrastructure and transportation, and increased employment flexibility, they are no longer as disconnected and isolated as they have been in the past. It should be possible to provide appropriate infrastructure, financial and professional support and adequate resources in remote regions. As human and technological advances continue, the remote workforce should reap the benefits through improved connectivity and endless innovation possibilities. The gap between the city and the bush is closing in many ways, it is imperative that one of these improvements is access to health services.

🧰 Manager's Toolkit

There are times when you will need to explain the unique geographical differences of your workplace to city-based managers. Take some time to think about how you will do this and write a short paragraph explaining the differences. Think about the best way to get the message across in a positive manner. Remember that they may not have a frame of reference possibly having never worked in a remote setting so try to be clear yet empathic. You may like to talk to you team about it at a team meeting and together develop a consistent approach to describing your remote workplace to outsiders. It may take some time to get the words as you want them, but once it is done you can use them over and over again (e.g. reports, recruitment, management meetings, requesting resources, etc.).

References

Baker T (2014) Attracting and retining talent: becoming an employer of choice. Palgrave McMillan, Hampshire, UK

Bent A (1999) Allied health in Central Australia: challenges and rewards in remote area practice. Aust J Physiotherapy 45(3):203–212

Buchan JM, Naccarella L. Brooks PM (2011) Is health workforce sustainability in Australia and New Zealand a realistic policy goal? Australian Health Rev 35(2):152–155

Buykx P, Humphreys J, Wakerman J, Pashen D (2010) Systematic review of effective retention incentives for health workers in rural and remote areas: towards evidence-based policy. Aust J Rural Health 18:102–109

Devine S (2006) Perceptions of occupational therapists practising in rural Australia: a graduate perspective. Aust Occupational Therapy J 53(3):205–209

Fisher KR, Fraser JD (2010) Rural health career pathways: research themes in recruitment and retention. Aust Health Rev 34(3):292–296

Garnett S, Coe K, Golebiowska K, Walsh H, Zander K, Guthridge S. Li S, Malyon R (2008) Attracting and keeping nursing professionals in an environment of chronic labour shortage: a study of mobility among nurses and midwives in the Northern Territory of Australia. Charles Darwin University Press, Darwin http://digitallibrary.health.nt.gov.au/dspace/bitstream/10137/228/1/nurse_report.pdf. Accessed 5 Sept 2017

Hays R, Wynd S, Veitch C, Crossland L (2003) Getting the balance right? GPs who chose to stay in rural practice. Aust J Rural Health 11(4):193–198

Hegney D, McCarthy A, Rogers-Clark C, Gorman D (2002a) Retaining rural and remote area nurses. The Queensland, Australia experience. J Nursing Adm 32(3):128–135

Hegney D, McCarthy A, Rogers-Clark C, Gorman D (2002b) Why nurses are resigning from rural and remote Queensland health facilities. Collegian 9(2):33–39

Humphreys JS, Jones MP, Jones JA, Mara PR (2002) Workforce retention in rural and remote Australia: determining the factors that influence length of practice. Medical J Aust 176 (10):472–476

Humphreys JS, Wakerman J, Wells R (2006) What do we mean by sustainable rural health services? Implications for rural health research. Aust J Rural Health 14(1):33–35. https://doi.org/10.1111/j.1440-1534.2006.00750.x

Kossek EE, Valcour M, Lirio P (2014) The sustainable work force: organizational strategies for promoting work-life balance and wellbeing. In Chen PY, Cooper CL (eds) Work and wellbeing: wellbeing: a complete reference guide, vol III, Wiley, pp 295–319

Kruger E, Tennant M (2005) Oral health workforce in rural and remote Western Australia: practice perceptions. Aust J Rural Health 13:321–326

Onnis L (2016) What is a sustainable remote health workforce? People, practice and place. Rural Remote Health 16:3806

Onnis L, Dyer G (2017) Maintaining Hope. J Mental Health Training, Edu Practice 12(1):13–23

Onnis L, Pryce J (2016) Health professionals working in remote Australia: a review of the literature. Asia Pacific J Hum Res 54:32–56

O'Toole K, Schoo AM (2010) Retention policies for allied health professionals in rural areas: a survey of private practitioners. Rural Remote Health 10:1331

Santhanam R, Hunter E, Wilkinson Y, Whiteford H, McEwan A (2006) Care, community, capacity: rethinking mental health services in remote indigenous settings. Aust J Primary Health 12(2):51–56

Wakerman J, Humphreys J (2012) Sustainable workforce and sustainable health systems for rural and remote Australia. Med J Aust 199(5):14–17

World Health Organisation (WHO) (2010) Increasing access to health workers in remote and rural areas through improved retention.WHO Press, France. http://www.searo.who.int/nepal/mediacentre/2010_increasing_access_to_health_workers_in_remote_and_rural_areas.pdf. Accessed 5 Sept 2017

Part II
Human Resource Management Challenges

Recruitment: Attraction, Advertising and Realistic Recruitment

4

When people are acting oddly, or differently from the way you'd act, ask yourself 'why?' We all react differently to various situations. What stresses out one person no end will not faze someone else in the slightest. Don't try to change the person; just try to understand why they are not behaving the way you expected.

Rachael Robertson, Antarctic Expedition Leader.
(Robertson 2014, p. 103)

Key Messages

- Recruitment advertisements communicate HRM policies to potential employees.
- The messages contained in recruitment advertisements and the recruitment process contributes to psychological contract formation.
- Recruitment advertisements attract potential employees; retention involves meeting their expectations.
- Remote health workforce sustainability could be improved through asking what attracted the current remote health workforce.
- Few advertised management positions had mandatory (or desirable) requirements for management qualifications, previous management experience or experience in a remote region.

When It Feels like Recruitment Is All You Do

'Many businesses and public sector organisations throughout Australia find it difficult to attract, let alone retain, staff. This 'problem' is exacerbated in remote and desert Australia, which is far removed from the attractions of the cities as well as the comprehensive infrastructure and services that are available in high population centres.'

McKenzie (2011, p. 354)

© Springer Nature Singapore Pte Ltd. 2019
L. Onnis, *HRM and Remote Health Workforce Sustainability*,
Management for Professionals, https://doi.org/10.1007/978-981-13-2059-0_4

In geographically remote areas, the 'war for talent' increases as organisations compete to attract and retain experienced, competent health professionals who also find the work and context both personally and professionally rewarding. The WHO (2010) predict a global shortage of health workers by 2035 with many countries already experiencing shortages of nurses and midwives. However, the maldistribution of health professionals, typically more in urban areas and a scarcity in remote regions creates additional workforce challenges (Campbell et al. 2013; WHO 2010). In fact, for many remote managers recruitment consumes a large proportion of their time. Therefore, effective recruitment practices are fundamental to improving workforce sustainability. The ongoing challenges of workforce attraction and retention are costing governments and organisations, in terms of time, money and lost opportunities (McKenzie 2011).

This situation suggests that in order to improve workforce sustainability, a multi-practice approach is vital. It is unlikely that one strategy alone will cut through the complexity of recruiting health professionals who embrace the challenges of working in remote health and enjoy living in remote regions. Current strategies include: recruiting health professionals with rural backgrounds; those who have completed work experience in rural and remote areas; prior to professional registration; and offering generous remuneration, employment conditions and financial incentives (Battye and McTaggart 2003; Hegney et al. 2002; Kent-Wilkinson et al. 2010; O'Toole and Schoo 2010). While, there is no doubt that these strategies are influencing attraction and retention; long-term reductions in turnover have not been widely reported. In fact, Hemphill and Kulik (2011, p. 117) found that despite the government strategies and incentives, there are declining GP to patient ratios in rural and remote Australia; concluding from their research that 'new recruitment strategies are needed.' Therefore, remote managers will need to adapt their current recruitment strategies so that they can attract a pool of qualified suitable applicants for (Sisodia and Chowdhary 2012).

Recruitment Advertising

An analysis of recruitment advertising was conducted in the study. The analysis showed the ways in which information is communicated about vacancies in remote regions. The appendix contains more information about how the recruitment advertisements were selected and analysed. In brief, over a two year period (August 2013–July 2015) online recruitment advertisements for health professionals, or positions managing health professionals, were reviewed and analysed. Over the two year period there were 3311 advertisements looking for health managers and health professionals to work in remote regions of tropical northern Australia. The majority of the advertisements were for nursing positions, more than half were full-time, and almost half were in Queensland (Onnis 2017).

As mentioned several times already, the global health workforce shortage is predicted to worsen, particularly in countries with ageing populations. In chapter one, labour mobility was introduced as a factor to be considered along with turn-over and workforce stability. With this in mind, it is important for remote managers to know the situation in their region, that is, the trends and labour shortages that are occurring where they are working. While it is good to know the global and national situation; it is important to know the local labour market too.

To explain a little about how the data is useful for recruitment, the findings from the study are used. The analysis of the recruitment advertisements showed that Northern Territory (Top End) and Western Australia (Kimberley) advertised fairly consistently although there was a slight increase for Northern Territory over the two year period. In contrast, Queensland (Far North and North West) had a considerable increase in advertisements during the study period. This suggests that for remote managers in the Northern Territory and Western Australia regions the situation remain consistent, it does not mean that the workforce was stable, it does not show whether turnover was low or high, it just shows that the amount of advertising was about the same across the two year period. In contrast Queensland had an a con-siderable increase in advertisements during the period, which suggests that there was more recruitment activity towards the end of the period. The remote manager should reflect on why they are seeing these trends. There are several reasons that can be explained by normal seasonal fluctuations, organisational policy changes, and labour movements. Some examples and explanations that may explain the changes in advertising patterns follow.

The increases can be explained by seasonal fluctuations such as:

- an increase in short-term contracts to cover annual leave taken towards the end of the year;
- natural attrition (people may prefer to leave at the end of the year);
- climatic changes (e.g. wet season in northern Australia is hot and humid);
- completion of work/training programs (e.g. annual 12 month placements);
- the completion of short-term contracts based on defined funding cycles.

The increases can be explained by organisational recruitment policies and practices:

- the flow of new employees into the organisation is slowed at times of financial uncertainly (e.g. the WA government website announced 'an immediate freeze on all recruitment until 30 June 2014 (unless otherwise approved)' (posted 16/4/2014).

- recruitment pools (e.g. as Western Australia Country Health Service (WACHS) had a range of regularly vacant positions (e.g. Registered Nurse, Enrolled Nurse, Remote Area Nurse, mental health clinician) listed in an ongoing online advertisement to be considered for a recruitment pool. This advertisement is continually attracting new applicants for the organisation.
- organisational restructures that result in retrenchments and redundancies often include policies that require organisations to recruit from within prior to seeking external applicants which can effect the number and type of positions advertised (i.e. if recruiting from within, external advertising is more likely to be for lower-level positions that become vacant as internal applicants are promoted.
- short-term funding cycles effect how organisations recruit and often lead to short-term or casual contracts that are a result of ongoing financial insecurity.
- organisational policy to only advertise on the organisation's website (e.g. government agencies), and not to use common recruitment websites (e.g. https:// www.seek.com.au/, https://careerone.com/)

These lists describe only a snapshot of the reasons that there may be observed changes in recruitment advertising over a 12 month period. These fluctuations and organisational policies may be able to explain the fluctuations; however, the trends and patterns may be revealing where and when there is real increased turnover in remote regions. Therefore, for remote managers, an awareness of workforce data can assist in understanding the patterns of turnover in the team, the organisation and the region (Russell et al. 2012). This information contributes to the remote manager's understanding of the recruitment strategies required for their team.

⑪ Reflection

In your reflective journal consider the following questions and answer the ones that you think are relevant to your situation.

1. How can knowing about fluctuations in recruitment advertising help me in my role?
2. If there is stability in the region, yet high turnover in the organisation how should my recruitment strategy be adapted?
3. If there is stability in the organisation, yet high turnover in my team, could I learn from the recruitment practices of other managers in my organisation who have better workforce stability?
4. This is irrelevant to me as I have high stability in my team, people rarely leave; however, I know that there is high turnover in other parts of the organisation. How can I assist other managers to stabilise their turnover for the overall benefit of the organisation and community?
5. How could we work together to create better workforce sustainability for the organisation?

Psychological Contracts

Psychological Contract Theory describes an individual employee's beliefs about 'what they think they are entitled to receive because of real or perceived promises' from their employer (Bartlett 2001, p. 337). An employee's belief about the reciprocal exchange agreement that forms the employment relationship is unspecified and implicit. Therefore, the beliefs that form the psychological contract are not known to anyone other than the employee (Cullinane and Dundon 2006). Consequently, remote managers are unaware of their perceived obligations. Despite being unaware of the obligations, employees expect that the remote manager will meet these obligations. There are a variety of actions, expectations and complications to the psychological contract beyond the obvious difficulties associated with an unwritten contract developed by only one party.

These obligations may be transactional (e.g. pay) or relational (e.g. loyalty in exchange for job security). Psychological contract formation commences during the recruitment process and it is during these early experiences that employees form expectations and perceived obligations (Knights and Kennedy 2005). Psychological contracts continue to be shaped throughout the employment relationship (Cullinane and Dundon 2006; Zhao et al. 2007). Where employees perceive that organisations meet their perceived obligations, they will reciprocate with positive attitudes and behaviours (Gould-Williams and Davies 2005). However, if an employee has unmet expectations, they may believe that the organisation has breached or violated their psychological contract (Knights and Kennedy 2005). A breach is described as the cognitive evaluation, that is, a mental calculation of what has been received compared to what the employee believes was promised, whereas, violation is the emotional response that may follow from the breach (Knights and Kennedy 2005; Zhao et al. 2007). In other words, a violation is the outcome of a perceived breach, and the emotion associated with the violation is often translated into the behaviour that results in voluntary turnover (O'Donohue and Nelson 2007; Zhao et al. 2007). For example, an employee may have perceived that the manager is obligated to provide a flexible roster so that they can spend the afternoons fishing because it was written in the recruitment advertising. However, the roster and on call responsibilities leave very little time for a relaxed lifestyle and fishing. They may be disappointed but accept the breach. Then, if a new employee arrives and is given a flexible roster and the first employee sees the new employee going fishing regularly; they may feel differently about the breach of their psychological contract and they may be quite emotional about this unfulfilled obligation. This emotion leads to the employee feeling that their psychological contract was violated because they upheld their part of the perceived exchange by working the inflexible roster. Inevitably psychological contract violation leads to turnover. The remote manager may not know exactly why the employee left so abruptly. The impact of seemingly trivial actions on employees' psychological contracts is an area of influence that should be considered by remote managers. Further information about psychological contract theory can be found in Appendix C.

⑪ Reflection

Take a moment to reflect on your own psychological contract. What are the unwritten expectations that you have of your employer? Think about the expectations that you have that are not written anywhere. Where did they come from? How did you develop these expectations?

Communicating with Potential Employees

Recruitment advertisements are not just a means of conveying information about the ideal person that you require for a vacant position; they communicate information about your organisation to potential applicants about the suitability of the organisation for them and their career aspirations (Green and Dalton 2007). Recruitment advertising promotes the employer's brand to the world. For successful applicants, the advertisement can be the commencement of the employment relationship, as such, it is here that psychological contract formation commences. Realistic recruitment advertisements can benefit organisations experiencing high turnover if employees commence the employment relationship with a more realistic image of the organisation, the role and the remote context.

Unrealistic recruitment advertising creates unrealistic employee expectations. Hence, psychological contract violation may be a contributing factor for high turnover and workforce instability in some remote regions. Generally, recruitment advertising included a brief overview of the organisation, the environment and the remote region. Frequently they described idyllic romanticised locations with sunsets, tropical views, red earth and warm climates, with only a handful providing less idealistic descriptions, such as, 'hot and humid', 'much of the areas you will be travelling into are on unsealed roads' and 'You will be isolated, you will be hot!' (Onnis 2017).

More than half of the recruitment advertisements in the study contained information about remuneration and incentives. The large incentive payments offered in recruitment advertising sends a message to potential applicants about the remuneration and incentives that are on offer, contributing to psychological contract formation. How these perceived obligations are formed is not difficult to imagine when the recruitment advertising contains statements such as, 'In addition to the great salary our employees enjoy an amazing range of benefits', 'Attractive pay rates and other incentives', and 'Fantastic way to save $$$ and experience life in stunning outback locations!!' (Onnis 2017).

Other factors associated with geographically remote work promoted in the recruitment advertisements included 'adventure' which attempts to attract those interested in the adventurous aspects of working in remote regions, e.g. 'Calling Rural Superheroes … To be successful you must be an all-round experienced generalist; fearless, flexible and ready to wear your red undies on the outside superhero style.' These advertisements, together with the ones using humorous catch phrases, may be viewed as under valuing the high levels of clinical competence required to work in remote regions and the endless hours of flying or driving on long hot dusty roads between clinics and clients. This study found that recruitment agencies were more likely to use this type of language in their advertising, suggesting that they were more actively using marketing approaches to attract potential clients, e.g. 'Midwives are as HOT AS BURNT TOAST in Australia and we simply CANNOT get enough of them' (Onnis 2017). In addition, it was only recruitment agencies that offered incentives like an opportunity to win a free iPad for a successful referral (Onnis 2017).

Many recruitment advertisements contained the organisation's values, vision and mission which is beneficial in attracting health professionals with values congruent with those of the organisation (Green and Dalton 2007). However, organisations recruiting through agencies may miss an opportunity to communicate their employer brand. Many advertisements in the study were agency branded, thus communicating the recruitment agency's values, mission and vision (Baum and Kabst 2014). This is important because not only do most people seek jobs with employers whose values, traits and characteristics are perceived to be similar to their own; the values, missions and vision statements contribute to psychological contract formation (Green and Dalton 2007; Lee et al. 2011).

In addition, the recruitment advertising often claimed to provide opportunities to 'make a difference' and described lifestyles that included leaving early to go fishing which creates expectations about the position and the employer. Employees who make decisions based on these descriptions may create a relational psychological contract with their employer where they anticipate a relaxed lifestyle of fishing and camping in exchange for providing the clinical services listed in the job description. If the employment experience results in the employee feeling that the employer has not provided their part of the arrangement, the employee may feel that the employer has breached or even violated the psychological contract (e.g. large workloads or 24 hours on-call responsibilities). As described earlier a breach of psychological contract precedes a perceived psychological contract violation and usually leads to voluntary turnover (Knights and Kennedy 2005; Zhao et al. 2007).

While much of the literature focuses on the employee's perception, it is essential to also consider the employer's perspective (Guest 1998). External market pressures, organisational structures, organisational culture, and institutional inertia may all contribute to a perceived psychological contract breach, so, if managers do not deliver on their perceived obligations, it may not be the manager's fault (Cullinane and Dundon 2006). However, remote managers can minimise the potential for such psychological contract breaches through transparent management practices and realistic recruitment practices. That is, where employees see that there has been a change in the organisation's situation they may be more accepting of a breach, provided the working relationship between the employee and their manager is sound.

Recruiting Managers

'I don't think we get too many people with management experience applying for management jobs …We usually recruit inexperienced managers more often than not and try to develop them and we don't do very well in developing them' HR Manager.' (Onnis 2014, p. 11)

Recruiting competent remote workforces includes recruiting competent managers suited to managing health professionals working in remote regions. As one HR Manager explained 'the complexity of providing health services in remote areas is challenging for managers, and particularly challenging for inexperienced managers who are isolated from professional development opportunities.' If accessing professional development and support in remote areas is difficult; this is not an ideal environment for an inexperienced manager, yet many managers thrive in remote areas often due to their personal characteristics, such as their previous management skills and experience, remote experience and personal resilience. These characteristics were the foundation on which they built their management career and contributed to the success that they had with improving workforce sustainability. There is no doubt that remote managers are vital, as one HR Manager proclaimed, 'Managers were key; they were the critical thing and they were the deal breakers' (Onnis 2014, p. 11).

There were 348 advertisements for management positions in the study, which accounted for 10.5% of the total advertisements over the two year period (Onnis 2016). Few advertisements mentioned management qualifications and/or previous management experience. In fact, an analysis of this subset of advertisements, found that management qualifications were mandatory in only 4% and desirable in only 9.5% of the advertisements (Onnis 2016). In addition, management experience was mandatory in 5.5% and desirable in 0.8% of the management advertisements (Onnis 2016). Moreover, only 6% of the management advertisements said that previous experience working in a remote region was mandatory or desirable (Onnis 2016). These findings are quite alarming, particularly given the challenges associated with accessing professional development and support discussed in the previous chapters, and the skills and experience required to work and manage health services in remote regions.

⊡ Reflection

Take a moment to reflect on your role as a manager, particularly the skills needed to be an effective manager in your remote location.

1. What are the most important skills needed?
2. How did you gain these skills?
3. How would you describe the level of experience needed to undertake your role?
4. How closely does what you do match the description in your position description?

Write a recruitment advertisement for your position, describe the role, list the duties and responsibilities, any specific qualifications needed, any qualifications that would be helpful (desirable) and what are the characteristics of a person who would thrive in your job? How would you describe the work environment, be honest (think about when you first arrived).

Manager's Toolkit

Take the recruitment advertisement that you wrote and put it in your toolkit. You can use this activity to shape your approach to future recruitment activities.

- The next time you are recruiting have a look at it and think about how you are describing the work environment.
- If you are applying for a new position, use it to concisely summarise what you do in your current role
- When you get that promotion, you have an advertisement ready to go for your old position.

Transition from Clinician to Manager

'How they got to be there was not necessarily because of their managerial and leadership skills. They got to be there because they were very good at their technical skills ... what happens is the wrong people go into the manager's job ... so we don't choose the right people very well; and the next thing is so we don't groom the next lot of people well. So we are not training the next generation of potential managers' HR Manager.

(Onnis 2014, p. 6)

The transition from clinician to manager takes many forms, and regardless of whether it is the result of a calculated career progression or whether it eventuates from a case of being in the right place at the right time; it is a significant career move for most health professionals. In the study, when managers remembered their first management position, most recalled an absence of support, with one saying, 'they knew me as a clinician, they all knew my background' (Onnis 2014). This manager felt that even though senior management were aware of her lack of skills and experience as a manager, they just expected her to commence in a management role and know what to do (Onnis 2014). This is not uncommon with other studies discussing these types of unsupported transitions saying that '[i]nadequate preparation of operational managers' and 'inadequate recognition of health services management as a health discipline' further impedes retention (Lenthall et al. 2009, p. 210). As such, organisations need to prepare the health professionals that they are promoting and proactively identify, train and support potential managers (Taylor et al. 2010).

While there are many reasons for this absence of support, one reason, emphasised by one manager was a lack of awareness of the type of support that they needed, saying, 'how do you know what you need because you don't know, what you don't know' (Onnis 2014). Similarly, the HR Managers said they were unaware of how well the remote managers were being supported through the systems that existed in their organisation. One HR manager explains how a misunderstanding about the level of support health managers receive can occur (see In Practice Box 4.1: George's story).

⒤ Reflection

How do you know what you don't know? Take a moment to reflect on your own experience in your first management role.

1. What was your experience?
2. How did you learn the things that you needed to know to be an effective manager?
3. Which strategies helped you and which ones did not work?

 Resources

Several organisations have created competency frameworks to assist managers and health services to develop the skills and experience deemed necessary to be a competent manager.

Australasian College of Health Service Management—Health Service Managers Competency Framework. https://achsm.org.au/Public/Public/Education_/Competency_ framework.aspx

Leadership Competencies for Health Services Managers by the International Hospital Federation.   https://www.ihf-fih.org/resources/pdf/Leadership_Competencies_for_ Healthcare_Services_Managers.pdf

American Organization of Nurse Executives. AONE Nurse Manager Competencies. Chicago, USA. www.aone.org Accessible at: http://www.aone.org/ resources/nurse-leader-competencies.shtml

The Clinical Leadership Competency Framework by the National Health Service (NHS) Leadership Academy, UK. https://www.leadershipacademy.nhs.uk/wp-content/uploads/2012/11/NHSLeadership-Leadership-Framework-Clinical-Leadership-Competency-Framework-CLCF.pdf

▦ Manager's Toolkit

A skills gap analysis is a helpful tool to identify areas where the skills and experience of the person compare to the skills and experiences required for the position. A skills gap analysis can be conducted by a person wanting to improve their own chances of career progression; by recruiters trying to assess the capabilities of applicants to the positions they are seeking; and by managers to identify where to focus professional development efforts for their current and future workforces.

Look at the list of skills, activities and responsibilities that you compiled in the reflection activity earlier in this chapter (or write a list now of the skills and experience you use in your current role). Write these skills in the left-hand column. Then, look at the position description for your current managers position or a position that you would consider to be a positive career move for yourself and write these skills across the top. Colour the boxes where you have the skills needed. Any columns without a coloured box are areas where you need to develop skills and/or experience to prepare yourself for the desired career move.

	Skill for desired role	Skill for desired role	Skill for desired role	Skill for desired role	Skill for desired role	Skill for desired role	Skill for desired role
Current skill		▓					
Current skill	▓						
Current skill					▓		
Current experience							
Current experience							

You can do this exercise with your team to help them to develop any skills that they may need for their current position, and also for career planning. It can be used for succession planning—so you can start to build capacity in your team to move into a remote manager role which is a sensible move for the sustainability of both your workforce and the remote health workforce more broadly. This will also help you to see the skill mix in your team and to identify areas where skills are lacking.

	Team member A	Team member B	Team member C	Team member D	Team member E
Skills needed to be a remote manager	▓			▓	
Skills needed to be a remote manager					
Skills needed to be a remote manager	▓		▓		
Skills needed to be a remote manager			▓		
Skills needed to be a remote manager		▓			

You can use the skills gap analysis for your own development and can draw on it when you apply for your next promotion. There are many variations to the template available for free so do an internet search until you find a style that suits you. Then, customise the template to suit your needs and put a copy in your Manager's Toolkit.

It is common for remote managers to have both clinical and management responsibilities. As clinicians, managers focus on client needs and this focus may, at times, influence their decision-making capabilities as managers. Several remote managers discussed their desire to maintain their clinical skills while in a management position, with one explaining it as 'their point of difference as a manager', saying, otherwise it would make more sense to appoint managers with management experience and qualifications (Onnis 2014, p. 10). While clinical skills may be their point of difference, it is important that they do not become a disadvantage. Hence, as managers, clinicians must ensure they develop management skills as well as maintaining their clinical relevance. Many managers transition from clinicians to managers in remote areas where both physical and financial barriers, contribute to a diminished level of access to management development and support. In the study, managers believed that if they had been in a metropolitan area during the transition into management they would have had increased access to management development and support which would have improved their capacity to manage more effectively (Onnis 2014). Only one manager identified a clear clinician to manager pathway, saying that they had progressed through the ranks from a registered nurse to management through a succession planning HRM policy (Onnis 2014).

The characteristics remote managers bring to the role (e.g. resilience, previous skills and experience) are the foundation on which they build their management career and contribute to the organisation's success. In the study, some managers provided examples about how they improved workforce sustainability, e.g. 'we had 51 vacancies across remote and then we got down to having a waitlist, so there was one time when we had every position filled with a permanent appointment and then we had a few people on the waitlist' and 'I took the turnover rate from 200% a year to basically we had one staff leave in the four and a half years that I was there' (Onnis 2014, p. 12). These managers showed that it is possible to transform health services with high turnover into health services with a stable workforce through management practices.

IN PRACTICE BOX 4.1: **When Support Systems are not Aligned with Management Roles and Responsibilities**

George's Story
As a HR Manager, I knew that we had systems in place to professionally support staff, including the managers. Then I found out that these systems did not support managers with their management skills. I had naïvely thought that the managerial/leadership aspects of their management role were part of their regular supervision conversation. So, as all the health managers had some form of profession supervision program (including the nurses), I had assumed that they were all getting the necessary support. I was wrong; nothing was in place for those managers.

Reflection
On reflection, it is not surprising that this is missing from supervision, as it was not viewed as part of their core responsibilities. Well not really, I mean how are the managers judged? How was their performance assessed? The managers are assessed on making budget and FTE, not on staff turnover, not retention, not staff satisfaction, none of the people management areas. So we didn't get the right people to start with, we weren't grooming the right people, we didn't put the right people in place, we didn't give those that we put in place the right support structure or development programs, and we never measured those people management/leadership aspects of the job anyway. So hey, it was serendipitous if you got something good out of it.

Action Taken
In response, the HR Manager set up formal supervision agreements with the managers for the non-clinical aspects of their positions with other senior

clinicians with management and leadership responsibilities or non-clinical managers with the experience in the management areas that had been identified as those where they lacked the necessary skills and experience. Participation was voluntary; it was for managers who wanted more support with their management development.

Attraction

While there are many reasons why health professionals choose to work in remote regions, the study found that there were eight reasons that could be used to attract future workforces (Table 4.1). The eight reasons were: Adventure/Travel; Autonomy; Geography; Indigenous Health; Lifestyle; Make a Difference; Remuneration; and Scope of Practice. These eight reasons were then used to conduct a content analysis of the recruitment advertisements (i.e. code the text of the advertisements to identify the frequency with which certain words/phrases are used). The analysis was conducted to determine whether the eight reasons that the current remote health workforce identified as being the factors that attracted them to work in a remote area were used in the recruitment advertising to attract applicants. The analysis of the recruitment advertising found that it does use these eight reasons to differing degrees and that there is opportunity to use them more (Table 4.1).

In Table 4.1 the column headed 'health professionals' shows the proportion of health professionals in the study that identified each reason. In other words, 8% of the health professionals who completed the question on the questionnaire said that they choose to work in a remote health service because they were looking for adventure or wanted to travel; 4% said that they wanted autonomy, and so on down the list. The column headed 'recruitment advertisements' shows the proportion of recruitment advertisements that contained that information. So, 8% of the

Table 4.1 A comparison of the reasons current remote health professionals choose to work in remote regions (n = 216) and the frequency in which these reasons are contained in recruitment advertisements (n = 3311) (Onnis 2017, p. 31, reproduced with permission)

Reason	Health professionals (%)	Recruitment advertisements (%)
Adventure/travel	8	8
Autonomy	4	2
Geography	1	14
Indigenous health	11	6
Lifestyle	18	8
Make a difference	6	6
Remuneration	4	56
Scope of practice	17	2

advertisements tried to attract potential applications by talking about remote health as an adventure or an opportunity to travel, 2% mentioned the autonomy of working in remote regions, and so on down the list.

This table highlights where there are opportunities to improve the way recruitment advertising targets potential applicants. For example, 17% of the health professionals in the study said that the 'scope of practice' attracted them to working in a remote region, yet only 2% of the advertisements included any information about the scope of practice (Onnis 2017). If organisations are advertising a vacancy that has a desirable scope of practice, the findings suggest that it is worth mentioning it in the advertising to attract the type of applicant who wants to work in a role that extends their scope of practice from the type of work that they have the opportunity to do in urban health services. Similarly, as 18% of participants in the study were attracted to 'lifestyle'. Therefore, if the remote location offers a desirable lifestyle, then including information about lifestyle in the recruitment advertising may improve the applicant pool. However, this comes with a caution: if the location does not offer a desirable lifestyle, putting that it does in the advertisement may improve attraction, but is unlikely to improve retention. Thus, make sure that the advertising is true to the experience which is discussed more in the next section about realistic recruitment.

Realistic Recruitment

'There are always people who are going to be wanting to go out there and work in this really amazing complex unique environment, that's not the challenge. The recruitment isn't the challenge it's the retention, and the retention is about management' Health Manager. (Onnis 2014, p. 7)

Recruitment strategies providing more realistic job previews are encouraged, particularly where applicants may be unfamiliar with the work context. In addition to realistic job previews, health professionals value realistic non-work information (e.g. local community and living conditions) (Richardson et al. 2008). Using realistic recruitment strategies to attract health professionals is beneficial for organisations experiencing high turnover because the new employee commences the employment relationship with information about the organisation, the role and the remote context. As these factors all shape expectations, realistic recruitment practices can lead to the formation of more realistic psychological contracts.

There are six factors considered to be predictors of an applicant's attraction (hereafter called 'attraction factors') to a particular job (Chapman et al. 2005):

- *Job and organisational characteristics*—applicants make choices based on the job attributes and organisational characteristics available to them. For example, communicating characteristics about diversity may attract a more diverse international and Indigenous applicant pool. Advertisements that provide realistic descriptions about the job advertised are more likely to 'generate a pool of qualified suitable applicants' (Chapman et al. 2005; Sisodia and Chowdhary 2012, p. 81);

- *Recruiter characteristics*—describes the influence of recruiters on the attractiveness of vacancies (e.g. friendly recruitment panel, membership of panel reflects diversity of applicants (e.g. gender, ethnicity, age) accommodating recruiter, inflexible interview schedule with fixed date and time);
- *Perceptions of the recruitment process*—focuses more specifically on the recruitment process, for example, procedural fairness, inclusivity (Chapman et al. 2005);
- *Perceived fit*—describes the subjective factors about the organisation or role that applicants seek. For example, recruitment advertisements provide the first impressions of the organisation and form a basis on which potential employees imagine the employment experience with the organisation; influencing recruitment because people join and remain with organisations when they perceive an alignment between the organisation's values and culture and their own values and aspirations (e.g. diversity, ethical business practices) (Chapman et al. 2005; Green and Dalton 2007);
- *Perceived alternatives*—describes the extent to which potential applicants perceive that alternative employment alternatives are available (Chapman et al. 2005). The continued narrative about geographically remote regions experiencing high turnover may make it more difficult as potiential applicants may percieve that there are many alternative remote employment opportunities; and
- *Hiring expectancies*—describes the applicant's evaluation of how likely it is that they may be offered the role (Chapman et al. 2005) which, similar to perceived alternatives, may be influenced by the narrative about high turnover in remote regions, and the diversity of the workforce.

🧰 Manager's Toolkit

Create a table with these six factors down the lefthand side and two columns to write in. The next time you have a vacancy, or if you have a position that is frequently vacant, try mapping the position against the criteria to determine the attractiveness of the position (see example below). This activity will help you to see why the position may not appeal to health professionals. The third column provides you with a list of the actions that you can take to increase the attractiveness of the position. Sometimes the actions that need to be taken are beyond your control, or are common to remote workplaces and not something that you can influence in your current position. This activity enables you to distinguish the actions that you can take and those that are out of your field of influence. As a remote manager, you can focus your efforts within your field of influence. Those beyond your immediate control, can be discussed with your manager.

Attraction Factors	Position (e.g. CNC)	Actions I can take
Job and organisational characteristics	List attributes of the job (highlight the ones that are unattractive to potential applicants (e.g. frequent oncall)	List actions you can take to improve the unattractive characteristics
Recruiter characteristics	Describe the recruiter's characteristics and how they influence applicants	List actions you can take so that the recruiter can improve the attractiveness of the position
Perceptions of the recruitment process	Describe the recruitment process (ask new staff so that you can understand it from their perspective)	List actions you can take to improve the recruitment process
Perceived fit	List the factors about the organisation or the position that applicants would find attractive	Match these factors to the ideal candidate to develop a description of the characteristics of the person with the best fit for the position
Perceived alternatives	List anything that makes this position distinctive, an opportunity not to be missed	List actions you can take to make the position appealing, and stand out from similar positions
Hiring expectancies	List factors that would influence how likely an applicant would get the role (e.g. has it been vacant for a long time?)	List actions that can you take to influence hiring expectations

If you find that this was a beneficial activity, turn it into a template and put it into your Manager's Toolkit so it is ready for you to use for another position in the future.

Local Recruitment

In the study, there were only a few health professionals who discussed the opportunity to recruit from local populations. The health professionals highlighted the social and economic benefits associated with employing a local workforce. Their argument was consistent with a report by Queensland Health (2017, p. 7) which stated that the 'training and employment of local health professionals correlates to longer tenure and less turnover.' Some participants discussed efforts to focus on developing career paths within their regions, such as 'grow your own' remote health professionals and managers. There is an increased focus on these types of programs, such as programs in rural and remote regions targeting high school and university students to create employment pathways for local communities and rural towns (Queensland Health 2017). The benefits are numerous, with health professionals ir. the study explaining that local residents know the region and are aware of working conditions, housing issues, resourcing inadequacies, local cultures and the environment so there are definitely benefits in employing locally.

IN PRACTICE BOX 4.2: They Could Have Prepared Me Better

Anne's Story

How it all Started

The first locum I did was in November 2001. I went in fear and trembling but I enjoyed it so much more than I expected. So for 12 months I did locums, three weeks on and one week off, which gave me a chance to come back to the city each month. During that time I got interested, because I would be working with these nurses who would be living out bush for a year and I'd think gosh, how does anyone do that? I thought, 'I would just like the challenge of being able to do that.' So as I went around doing locums I would think of things like: How do you get your hair cut? How do you do your banking? How do you do grocery shopping? And I'd ask the nurses how they do all those things and after about 10 months, I thought right, I am going to work permanently out bush. So, low and behold, I did a locum in a remote Aboriginal Community and loved it. It was the best locum that I ever did, and a few months after I was there the position became vacant and I was given the position—so I went for a locum and I ended up staying there three years!

I left there exhausted, but feeling like I had contributed something, if not very much, something. It was a wonderful cultural experience and I knew how I'd do living out bush for extended periods of time. I wasn't really planning on going back until I received a phone call from someone saying we just looked on our roster and in a fortnight's time we need a locum and the boss said to ring you because you can hit the ground running. I was so delighted that they thought of me like that and I did that locum for them and I have been doing locums back there ever since.

That One Big Worry, Which Could Have Been Prevented

When I went on that first locum, I was expecting lots of crime, I thought I might be raped, I expected not to have satisfactory accommodation and I expected that my accommodation would be broken into. I had many concerns, and one very specific concern. When I got all the paperwork from the recruitment agency that I had to fill in to go bush, even for the first short stint, I had to apply for a permit through the Aboriginal Land Council. I had to tick a box to say which town office I was going to pick the permit up from but I had no idea which office I would be picking it up from and I just didn't know the ropes. I thought I am going to have to trust someone from the agency here, I don't remember which box I ticked but the day came and I was on the plane, I was on my way and I didn't have the permit in my hand. All the paperwork that came with the application for the permit made it very clear you did not try and tempt fate by landing yourself into a remote community without having this permit on you. And I was going to be landing on a commercial plane and I did not have this permit on me and I did not know what would happen but I thought, maybe there will be a dozen Aboriginal warriors to meet the plane to check that as you come off the plane that you can show your permit and I didn't have the permit to show. I thought how am I going to explain myself to people who don't know me? I walked off the plane and into the terminal, there were no Aboriginal warriors, in fact no-one approached me. In this remote community, I had a very good quality house, I was not broken into, and I was not raped. My expectations were all far from the truth of how it really was there, and I realised that I was the one that needed to be educated about remote communities.

Did You Have all the Resources You Needed?
The agency could have helped me to know more about what to expect in a remote community. There are some things that you have to just experience for yourself to really understand but they could have put my mind at rest about some specific concerns. They could have been more thorough about the permit and the paperwork, I mean they were the ones sending my paperwork to the Aboriginal Land Council office, why didn't they go and pick up the permit for me. There would have been plenty of time, they could have posted it to me and I could have taken it with me. It would have prevented a lot of anxiety. So I learnt through experience about how things worked. They could have prepared me better and they didn't. Life in the city is so different from life in rural and remote Australia.

Expectations

'The work is so much more than I expected. The bits that met my expectations were travelling, working with lots of amazing people, getting a lot of experience. I didn't expect that I would fall in love with this work and not be able to imagine working anywhere else. I also did not expect that the lifestyle I have developed would be so fulfilling and meaningful to me as a person.' (Onnis 2016, p. 120)

The study found that for about half (53%) of the health professionals the work was as they expected, and a further 5% were non-committal reporting that some aspects were as expected and others were not as expected. In contrast, a large proportion (42%) found that the work was not as expected. For example, one health professional said 'Intellectually I knew what to expect but as always the actual exposure is challenging initially.' Other health professionals reported that the lifestyle did not meet their expectations, saying that 'it can sometimes be difficult to achieve work/life balance due to the amount of travel and the level of exhaustion after a trip', another said that 'it can get a little isolated sometimes.' As mentioned in the previous section, realistic recruitment practices, such as clear descriptions of the actual job are important for improving the match or person-fit, and in finding the right health professional to suit the position. One participant explained that they were not aware of the personal characteristics necessary to adapt their new work (and living) environment, saying, 'Initially, I thought it would be a "walk in the park" compared to working in NSW metro. How wrong was I—I never worked this hard in Metro.'

When considering expectation it is also worth considering motivation. In an effort to understand the motivation of health professionals for very remote work, Tyrrell (2017) developed a set of motivation scales. Tyrrell (2017, p. 275) developed the scales to determine the 'nature of the predominant motivations associated with choosing the very remote Indigenous community workplace' as well as helping to assess the likelihood of a substantial length of time working in the remote workplace. Tyrrell (2017) proposes that his predictive model will have benefits if used to assess potential applicants for very remote work, especially, but not limited to very remote communities. While it is unlikely that a remote manager would base recruitment decisions solely of this type of psychological assessment tool; it could be helpful in decision-making, particularly when recruiting health professionals without previous remote work experience. Further research in this area, particularly the refinement and development of psychological screening tools could lead to improvements in the recruitment process that can assist remote managers to develop management practices when selecting health professionals and ultimately, will improve workforce sustainability. For remote managers before using psychological tests in recruitment it is recommended that you contact your HR Department to find out which psychological tests are available and appropriate for your circumstances (this may differ by state/territory and some tests can only be administered by qualified psychologists).

Effective Recruitment Practices

'I have always worked remotely all my working life because I choose to. It is more difficult with less support so things like orientation and helping people start-up are all the more important. I don't think employers get this right too often. They do not spend enough time initially on getting someone comfortable with their new role' Remote Health Professional.'
(Onnis 2016, p. 116)

The information contained in the advertisements includes many HRM policies (e.g. salaries, leave entitlements). The difference between the HRM policies communicated in recruitment advertising and the conditions experienced by the health professional in remote regions contributes to the formation, perceived breaches (both relational and transactional) and possible violation of the psychological contract. Hence, the information contained in recruitment advertisements influences the sustainability of remote health workforces.

More focused recruitment practices that seek to attract health professionals using the factors that health professionals currently working in remote regions find attractive is a sensible recruitment strategy. Hence, incorporating the 'attraction factors' described by current health professionals in recruitment advertising will be beneficial for recruitment. However, the study findings suggest that improvements for health workforce sustainability should be focused on the management practices around interpretation, implementation and localisation of HRM policies in remote workplaces. As such, recruitment is only one component of an integrated HRM approach to remote health workforce sustainability. In a geographically remote context, management practices that value the experience of current employees and provide strategic HRM solutions to workforce challenges are essential. Hence, tailoring recruitment practices to the remote context improves workforce sustainability, which in geographically remote regions improves access to healthcare services.

References

Bartlett KR (2001) The relationship between training and organizational commitment: a study in the health care field. Hum Res Develop Q 12(4):335–352

Battye KM, McTaggart K (2003) Development of a model for sustainable delivery of outreach allied health services to remote north-west Queensland Australia. Rural Remote Health 3:194

Baum M, Kabst R (2014) The Effectiveness of recruitment advertisements and recruitment websites: indirect and interactive effects on applicant attraction. Hum Res Manage 53(3):353–378

Campbell J, Dussault G, Buchan J, Pozo-Martin F, Guerra Arias M, Leone C, Siyam A, Cometto G (2013) A universal truth: no health without a workforce. WHO Press, Geneva http://www.who.int/workforcealliance/knowledge/resources/GHWA_AUniversalTruthReport. pdfAccessed. Accessed 3 Sept 2017

Chapman DS, Uggerslev KL, Carroll SA, Piasentin KA, Jones DA (2005) Applicant attraction to organizations and job choice: a meta-analytic review of the correlates of recruiting outcomes. J Appl Psychol 90(5):928–944

Cullinane N, Dundon T (2006) The psychological contract: a critical review. Int J Manage Rev 8 (2):113–129

Gould-Williams J, Davies F (2005) Using social exchange theory to predict the effects of HRM practice on employee outcomes. Public Manage Rev 7(1):1–24

Green J, Dalton B (2007) Values and virtues or qualifications and experience? An analysis of non-profit recruitment advertising in Australia. Employment Relations Record 7(2)

Guest DE (1998) Is the psychological contract worth taking seriously? J Organizat Behav 19 (S1):649–664

Hegney D, McCarthy A, Rogers-Clark C, Gorman D (2002) Why nurses are attracted to rural and remote practice. Aust J Rural Health 10(3):178–186

Hemphill E, Kulik C (2011) Segmenting a general practitioner market to improve recruitment outcomes. Aust Health Rev, 35(2): 117–123

Kent-Wilkinson A, Starr L, Dumanski S, Fleck J, LeFebvre A, Child A (2010) International nursing student exchange: rural and remote clinical experiences in Australia. J Agromedicine 15(1):58–65

Knights JA, Kennedy BJ (2005) Psychological contract violation: impacts on job satisfaction and organizational commitment among Australian senior public servants. App HRM Res 10(2):57–72

Lee C, Hwang F, Wang M, Chen P (2011) Hype matters applicant attraction: study on type of publicity and recruitment advertising. African J Business Manage 5(7):2734–2741

Lenthall S, Wakerman J, Opie T, Dollard M, Dunn S, Knight S, MacLeod M, Watson C (2009) What stresses remote area nurses? Current knowledge and future action. Aust J Rural Health 17 (4):208–213

McKenzie FH (2011) Attracting and retaining skilled and professional staff in remote locations of Australia. Rangeland J 33(4):353–363

O'Donohue W, Nelson L (2007) Let's be professional about this: ideology and the psychological contracts of registered nurses. J Nursing Manage 15(5):547–555

Onnis L (2014) Managers are the key to workforce stability: an HRM approach towards improving retention of health professionals in remote northern Australia. In: Proceedings, 28th australia and new zealand academy of management (anzam) conference. UTS Sydney

Onnis L (2016) A sustainable remote health workforce: translating hrm policy into practice. Ph. D. Thesis. James Cook University

Onnis L (2017) Attracting future health workforces in geographically remote regions: perspectives from current remote health professionals. Asia Pacific J Health Manage 12(2):25–33

O'Toole K, Schoo AM (2010) Retention policies for allied health professionals in rural areas: a survey of private practitioners. Rural Remote Health 10:1331

Queensland Health (2017) Advancing rural and remote health service delivery through workforce: A strategy for Queensland 2017–2020, State of Queensland (Queensland Health), Brisbane. https:// www.health.qld.gov.au/__data/assets/pdf_file/0039/672978/rural-remote-workforce-strategy.pdf Accessed 2 June 2018

Richardson J, McBey K, McKenna S (2008) Integrating realistic job previews and realistic living conditions previews. Personnel Rev 37(5):490–508

Robertson R (2014) Leading on the Edge. Wiley, Australia

Russell D, Humphreys J, Wakerman J (2012) How best to measure health workforce turnover and retention: Five key metrics. Australian Health Rev 36(3):290–295

Sisodia S, Chowdhary N (2012) Use of illustrations in recruitment advertising by service companies. J Ser Res 12(2):81–109

Taylor R, Blake B, Claudio F (2010) An analysis of the opinions and experiences of Australians involved in health aspects of disaster response overseas to enhance effectiveness of humanitarian assistance. University of Queensland, Queensland, Australia

Tyrrell M (2017) Health practitioner motivations in choosing the very remote indigenous community workplace: developing a scale to describe and measure them and their relationship to total length of stay. Ph.D. Thesis, Finders University

World Health Organisation (WHO) (2010) Increasing access to health workers in remote and rural areas through improved retention.WHO Press, France. http://www.searo.who.int/nepal/mediacentre/2010_increasing_access_to_health_workers_in_remote_and_rural_areas.pdf Accessed 5 Sept 2017

Zhao H, Wayne S, Glibkowski B, Bravo J (2007) The impact of psychological contract breach on work-related outcomes: a meta-analysis. Personnel Psychol 60(3):647–680

Remuneration: Extrinsic and Intrinsic Rewards, Incentives and Motivation

5

> *Rewarding an activity will get you more of it. Punishing an activity will get you less of it.*
> Daniel Pink, Author of several best-selling books on business and behaviour.
>
> (Pink 2009, p. 32)

Key Messages

- High remuneration and incentives attract people to remote regions but do not necessarily improve retention unless they are sufficient to prevent dissatisfaction.
- Health professionals have individual motivations; therefore, individual motivations will shape the employment relationship and influence turnover intentions
- In remote regions, workforce sustainability is contingent on both extrinsic and intrinsic rewards.
- High remuneration is viewed as a strategy to attract health professionals to remote areas, with a combination of extrinsic and intrinsic rewards viewed as vital to improve retention.
- In workplaces where there is high voluntary turnover and labour mobility, a combination of extrinsic and intrinsic rewards improves workplace sustainability.

Remuneration as Compensation

Remuneration describes the compensation received in exchange for working in remote regions. Remuneration includes monetary benefits, such as: pay, allowances, bonuses and financial subsidies. The primary focus for most organisations is to

© Springer Nature Singapore Pte Ltd. 2019
L. Onnis, *HRM and Remote Health Workforce Sustainability*,
Management for Professionals, https://doi.org/10.1007/978-981-13-2059-0_5

design a job so that employees work efficiently, receiving a level of remuneration commensurate to the work they do (Giancola 2011). Remuneration provides extrinsic rewards for people more extrinsically motivated.

Extrinsic motivation is where an activity is done in order to gain a reward that is separate to the activity (Ryan and Deci 2000). Pay is an extrinsic reward, as it is separate from the activity in that the only connection is that the reward is given if the activity is completed (Kanungo and Hartwick 1987; Giancola 2011). In contrast, intrinsic motivation is when someone does an activity for the inherent satisfaction they feel from the activity, in other words, the reward is the activity itself (Ryan and Deci 2000; Kanungo and Hartwick 1987). Intrinsic rewards include: promotion, authority, responsibility, and participation in decision making, praise from supervisors, co-worker recognition, and awards for excellence and superior performance (Giancola 2011). Economic principles suggest that people respond to incentives; however, these incentives, which are usually in the form of rewards and punishments, are often counterproductive and undermine intrinsic motivation (Gagné and Deci 2005; Ryan and Deci 2000). Further, while large financial incentives such as bonuses can encourage improved performance, these payments can make people dependent on the money, which diminishes any intrinsic motivation (Deci 1972). This could lead people to re-evaluate the activity transforming it from something intrinsically motivating to something done with the expectation of monetary reward.

Financial Incentives

'By offering a reward, a principal signals to the agent that the task is undesirable... There's no going back. Pay your son to take out the trash and you've pretty much guaranteed that the kid will never do it again for free.' (Pink 2009, p. 52)

The recruitment advertisements analysed in the study contained information about remuneration. Generally, the recruitment advertisements included descriptions of the remuneration offered, such as salaries, salary sacrificing, and other work-based incentives. In addition, there were benefits to compensate for geographical remoteness and the associated hardships of living and working in remote regions, including: remote allowances, return annual airfares to the nearest city (or a commensurate airfare); electricity or air conditioning subsidies to offset the costs of additional air conditioning in the warmer months; and, accommodation where there is limited access to private accommodation. In addition, there were specific retention incentives based on profession or geographic location. For example, some advertisements provided incentives that were *only* available to health professionals who relocated to the remote region. In contrast, others stated that there was no accommodation or relocation assistance available saying that 'Local residents only need apply.' These two vastly different approaches highlight the differences in reward systems between organisations, professions and geographic locations as well as available resources which is discussed further in Chap. 9.

Incentives

Many advertisements used a range of incentives to encourage health professionals to relocate to remote regions. The study identified 49 incentives mentioned in the advertising (Table 5.1).

Further analysis of the recruitment advertising found that while there were a variety of incentives, there were only eleven that were included in more that 1% of the advertisements (Table 5.2). Bonuses and financial incentives were mentioned in very few advertisements which is surprising given the rhetoric about how financial incentives are needed to attract health professionals to remote regions and the associated high financial costs for organisations.

The most frequently advertised financial incentives were above Award pay rates or a high salary, which is associated with the position, and therefore the health professional's skills and experience. Incentives not associated with the health professional's skills or expertise such as free uniforms, indemnity insurance reimbursements and the opportunity to travel were more frequently included in advertisements by recruitment agencies. The high frequency of advertisements by agencies for referral bonuses and free uniforms was interesting. The former suggests a more marketing focused recruitment policy where the aim is to collect details for potential candidates but only pay on successful placements. This policy entails a low risk strategy for the agency and a cost-effective way to source potential candidates. The latter, free uniforms, was more intriguing. Investigation, suggests that the uniforms usually displayed the name of the agency and/or the agency logo which is an effective advertising strategy for the agency. For the employee the benefit is in the value of the free uniform and any associated intrinsic benefit of the association with the company. While these are speculative suggestions about the benefits, the frequency in which they were mentioned in the advertisements suggests recruitment strategies in play that may influence many of the HRM concepts know to improve retention, such an organisational-identity. In other words, when a health professional wears a branded uniform they are clearly identified as an employee of that agency. This may influence their connection to the health service, their commitment to the health service's objectives, and their sense of organisational identity.

Table 5.1 **Remuneration and financial incentives from the recruitment advertising**

1	Annual isolation bonus	26	Indemnity insurance reimbursed
2	Bonus	27	Shift allowance
3	FOIL (Fares Out of Isolated Localities)	28	Badge
4	Gratuity payment (after 12 months/after 2 years)	29	Communications package (phone, laptop, internet access)
5	25% base rate in lieu of overtime/overtime allowance	30	Incentive package (doctors only)
6	QANTAS frequent flyer points	31	Free criminal history check
7	Uniforms (free/allowance)	32	Gym membership
8	Paid day off for birthday	33	Free dentistry
9	Travel allowance/travel incentives	34	Allowances (general)
10	Bonus at completion of placement	35	HECS reimbursement (33% for each year)
11	No weekend work	36	Additional leave
12	Referral bonus	37	Free car parking
13	Inaccessibility incentive	38	Public Holidays as paid days off
14	Private practice arrangements	39	Sponsorship
15	Isolation leave	40	Iconic organisation
16	Professional development/training	41	Attraction allowance
17	Financial incentives	42	Mobile phone allowance
18	Tax benefits	43	Petrol allowance (weekly)
19	Vehicle/vehicle allowance	44	High salary/above award rates
20	Managerial bonus/medical managers allowance	45	Incentive program
21	Availability allowance	46	Recreation facilities
22	Performance bonus	47	Disincentive (e.g. no meal Allowances)
23	Travel (opportunity to travel)	48	Corporate health plan
24	Utilities covered (or subsidised)/phone line	49	Spouse employment/assistance with finding employment for spouse
25	Subsidised meals		

Retention Incentives

Retention incentives differed considerably, with many nurses working in Queensland eligible for the Remote Area Nursing Incentive Package (RANIP) as outlined in the 'Nurses and Midwives (Queensland Health) Certified Agreement (EB9) (2017). The RANIP's annual isolation bonus is: Year 1 ($3500); Year 2 ($10,500); and Year 3 ($7000). The RANIP annual isolation bonuses are designed to improve retention in areas where turnover is high, with the higher amount incentivising the completion of two years in a remote region. A similar scheme is available for

Table 5.2 Most frequently advertised incentives

Incentives	%
High salary/above Award pay	13.07
Uniforms (free/allowance)	6.32
Professional development	4.20
Referral bonus	2.89
Financial incentives	2.18
Vehicle allowance	2.15
Bonus	2.02
Travel (opportunity/experience)	1.68
Indemnity insurance (reimbursement)	1.37
Annual isolation bonus	1.34
Incentive package[a] (doctors only)	1.12

[a]Private use of fully maintained vehicle, communications package (mobile phone, laptop, etc.), professional development allowance, professional development leave 3.6 weeks p.a., professional indemnity cover, private practice arrangements plus overtime and on-call allowance

medical professionals with the Australian Government's Department of Health (2017) publishing information on the *General Practice Rural Incentives Program website* (http://www.health.gov.au/). Medical practitioners are offered retention bonuses for working in remote areas and very remote areas, as follows:

	Year 1	Year 2	Year 3	Year 4	Year 5 plus
Remote	$16,000	$16,000	$25,000	$25,000	$35,000
Very remote	$25,000	$25,000	$35,000	$35,000	$60,000

From a HRM perspective these policies are designed to encourage retention and are paid based on the length of time the health professional remains in the remote location. These rewards are not performance based and, in fact, very few advertisements included any performance-based rewards. This can be challenging for remote managers as typically managers reward staff for excellence and good performance, whereas these incentives arbitrarily reward health professionals for staying regardless of the quality of their work or their workplace performance.

[!] Reflection

Think about your management style and how you manage your team do you use extrinsic or intrinsic incentives to motivate people? Consider your own motivation —is it intrinsic or extrinsic? Understanding yourself and your team can help you to adapt your management style to be more effective. For example, sometimes a heartfelt thank-you is more rewarding that a $10 voucher.

 **Manager's Toolkit**

Table 5.1 contains a list of 49 incentives that were offered to health professional in the recruitment advertising. Make a list of the incentives that are offered by your organisation. Are you using these incentives to attract the health professionals to your team?

Divide the list of incentives that you offer into intrinsic and extrinsic incentives— is this a good mix? In order to localise your management practices, think about your team and the type of benefits that they value.

1. Which team members are more extrinsically motivated, e.g. who will work additional hours if the pay is right?
2. Which team members are more intrinsically motivated, e.g. who values the time they spend with family so much that no amount of pay will entice them to work additional hours?

Create a list of incentives and rewards that you can offer to your staff to reward good performance. Some of the incentives will be consistent with your organisation's HRM policies, some will have no monetary value (e.g. thank-you, a coveted role at work) but they will all be considered a reward by your team. Put this list in your Manager's Toolkit and remember to add to it as you get new ideas and to use it to reward your staff.

Resources

In the book, *Drive—The surprising Truth About What Motivates Us* (2009), Daniel Pink introduces his ideas about motivation and how organisations need to rethink motivation. In this book Pink draws on research about intrinsic and extrinsic motivation, and provides practical examples from industry and education to explain motivation. The book also includes activities and suggested further reading if you are interested in finding out more about the topics covered. Daniel Pink has a regular newsletter, podcasts (called Pinkcasts) and an informative website with additional information for those interested in motivation and behaviour. https://www.danpink.com/.

Policy Conflicts

If an employee's expectations are formed prior or during the recruitment process, it follows that they are more likely to have a positive employment experience if their expectations are met. Further, realised expectations are more likely to translate into improved retention. Therefore, if there is a difference between the employee's expectations and their perceived reality, there is an opportunity for a psychological contract breach or violation as explained in chapter four. Some of the incentives offered could potentially lead to differences in perceived obligations and what health professionals experience in remote regions. For example, some advertisements stated that accommodation was negotiable, this may lead potential employees to expect that there is an opportunity to negotiate free or subsidised accommodation, yet when they arrive in the remote workplace, they may be advised that there is no free accommodation available or that they are not eligible for a subsidy in that particular region; hence, they may feel that the employer did not fulfil their perceived obligations, or even worse, they may feel that they were deceived. In this example organisational policies may contradict local policies, that is, the employee may be eligible under the organisation-wide policy; however, inadequate resources mean that the local manager is unable to implement the policy in the manner in which the organisation intended. Potential conflicts should be considered in the recruitment process to minimise the formation of unrealised expectations.

ⅱ Reflection

Remote managers often find themselves trying to implement policies that were developed in city-based head offices by senior managers who do not understand the remote environment sufficiently to foresee the possible policy implementation issues.

1. Do you have any policy conflicts in your workplace?
2. Think about your experiences localising policies so that they can be implemented in your remote workplace. How successfully were they implemented?
3. Is there room for improvement?
4. What strategies do you have in place to assist you to localise the implementation of policies?

Perceived Inequities

Perceived organisational inequities are considered a hindrance to recruitment and retention of remote health professionals. In the study, managers explained that the remuneration, incentives and working conditions that smaller health services and non-profit organisations can offer were often lower than those offered by government health services that had greater access to housing, facilities and alternative funding sources. While many of these inequities are evident in the entire health system; in remote regions they appear to be more pronounced as health professionals live as well as work in these inequitable conditions (see In Practice Box 5.1: Helen from HR's story).

The HRM policies that outline incentives and bonuses, particularly those that are prescribed in industrial agreements, such as enterprise agreements and Awards are transparent in structure. This means that all the employees are aware of the salaries and financial incentives that other employees are receiving. In small remote workplaces this can be difficult for managers who may have a small team who are receiving vastly different salaries and benefits (Hegney et al. 2002b; Santhanam et al. 2006). This is particularly difficult for multi-disciplinary teams where team members may be undertaking a similar role yet receiving different salaries, a different number of annual leave days and often different incentive payments. For example, in the study there were advertisements for a mental health clinician, a position that could be filled by either a nurse or an allied health professional. The successful candidate needed to be degree qualified and registered with a professional body, so the level of qualification and competence was similar; however, the role description showed that the starting salary for a nurse was $21,677 more than for an allied health professional. For this position, the nursing Award had four increment levels and the allied health Award had nine increments. If the successful applicant was an allied health professional they would not only begin on a lower pay rate, incrementally the Award pay increases never closed the gap, so that even after four years in the position there is still a difference in salary of $11,030 per year. It took seven years for the allied health position to be remunerated at a comparable rate (the fourth increment) for the nursing position: $83,760 (allied health) and $84,268 (nursing). However, an allied health professional must also have a post-graduate qualification to reach the highest increment further emphasising differences in expectations and benefits for different professions. In addition, there were different annual leave entitlements, different professional development provisions and nurses may be eligible for RANIP which included an isolation bonus, remote allowance, additional study leave and subsidised accommodation. All these differences in compensation were for the same position, with the same tasks, and the same level of responsibility. The differences were associated with the successful candidate's profession, not the role or their performance in the role.

IN PRACTICE BOX 5.1: **The Crux of the Challenge: Managing Pay Differences in the Workplace**

Helen from HR's story

Background
The crux of the challenge is that it is difficult to find health professionals to work in the remote areas so this organisation has the mentality that they will take whoever comes along. They are not setting the standards for who works for us; they take who they can find. They have the view that we are lucky to get someone who is skilled to go to the places that we ask them to go to, and so we will put up with whatever short comings they bring to the job. That is not communicated, but it is implied throughout the organisation. So what you end up with is people who are, and this has been said to me a few times, they are either a martyr, mercenary or a misfit. So they either feel that they are destined to go to wherever the job is and change that community in some fabulous way that only they can do, or they are extremely focused on the financial and other benefits and are very good at making that part of the equation work for their individual needs.

The issue

So the difficulty is that you often get a complete mismatch so you get team leaders who take too much accountability and do not consult and do not see themselves as being part of the corporate management team. Alternatively, you get other people who do not really understand leadership at all and they think that they are in the job because they are a good clinician. So they spend a lot of their time distracted by this professional barrier, for example, one team leader says that they cannot get through to a team member because she is a psychologist and they are a social worker and the disciplines are so different. But that is really the wrong conversation to be having if you are a manager. I am constantly astounded by that enormous skills gap for managers and I am often astounded at the behaviour of our team leaders. They do not go out there with any business training at all. They are really good in the community, and on the ground they do a fantastic job but they are lacking management skills. Recently, I had a situation where two health professionals were working in inequitable conditions but it did not come to my attention until after one had resigned. On their last day of work I happened to be visiting the remote community and she told me that she was doing the same job as the person working beside her and she was getting paid $30,000 less.

Actions

I followed up with the team leader and asked if she knew about it, and she did. When I asked what she had done about it, expecting her to say that she had spoken with her manager, and had taken some action to improve the inequity, all the team leader said was 'I told her, I made sure she knew that she was being paid $30,000 less'. That was what she thought was the appropriate action as a manager. It was because she was trying to be transparent and to embody the values about inclusive practice and consultation. These were not a relevant set of guidelines for the type of problem she was facing. It did not occur to her to contact HR, it did not occur to her to get any advice, and it did not occur to her that she was exposing the business to a lot of risk. Instead, the team leader thought it was OK as long as the person knew about it and was happy to continue to work within this inequitable situation.

Future directions

There is an enormous gap between the skills and experience that many health professionals have and those needed to effectively manage in remote communities; yet, these managers can do incredible things to retain, motivate, develop, and engage their staff. Every day they work in such hard conditions. While, it is a very big challenge to get people who care about compliance and best practice in business to go and work in these roles; it is essential for the wellbeing of the manager, business and the remote community. There is a need to consider contemporary understanding about the role of the manager. Often, line managers do not understand corporate capability models, and they have expectations that HR should be jumping on a plane and going to a work area every time there is an issue, rather than recognising that it is a line manager's job to manage their staff; and that corporate HR has the responsibility to set up the tools, the processes, and the coaching that managers need.

Motivation and Job Satisfaction

Remuneration is central to why people work; yet, despite increased financial incentives, there is little evidence about financial incentives improving workforce supply in rural and remote areas (Buykx et al. 2010). In fact, WHO (2010) suggest that retention incentives and interventions are more likely to be effective when the motivation to remain in remote regions comes from 'developing, deploying and motivating effective local service managers and strengthening human resources management systems (WHO 2010, p. 30). The WHO (2010) findings convey a message similar to the findings from the study in proposing that incentives and rewards in isolation are not effective in improving long-term workforce retention (WHO 2010). This may be why, when many health professionals reported job satisfaction; they reported satisfaction with specific aspects of their work not the entire employment experience. That is, they report satisfaction with aspects of their work as well as remuneration suggesting that satisfaction is achieved through intrinsic motivators as well as financial rewards.

The managers who participated in the study emphasised that intrinsic motivation is important for retention, with one manager saying, 'money and conditions is what will attract people, retention is around the work environment' (Onnis 2014). The managers suggested that the motivation to work in remote regions arises from either an intrinsic altruistic drive or extrinsic incentives. This supports the findings of other studies where extrinsic rewards were beneficial in attracting health professionals to remote areas; however, they had minimal impact on long-term retention (Campbell et al. 2012). One remote health professional explained,

'FIFO mining has proven that money does talk. If you pay people enough they will come and work here. That was a big reason for me coming out here and I guess if I wasn't remunerated well I wouldn't have continued to live here, especially after [the death of a child patient] 18 months ago. So I would have been out of here, definitely, because it's emotionally draining and just a, horrendous, horrendous day. So I guess you have to have the financial incentives in place and then the intrinsic stuff, well we don't have any control over that anyway. We were just lucky that my partner could get work and it's quite a nice community, it's not too violent, we haven't been broken into, those things as well.'

When the same health professional was then asked 'Do you think that even if the money was right, if things didn't meet your intrinsic needs that they would override the money or do you still think the money is enough to keep you there?' The remote health professional replied,

'I think, it depends on what community you are in. There are a lot of lovely little communities that aren't too busy and you could keep going. [This community] is not one of those communities regardless of how good the money is …Yes, sometimes, I think that money is a good way of getting people out here but it may not keep them here if they are not suited to it.'

Is this narrative it is evident that the financial rewards and incentives and intrinsic needs are closely intertwined and at times can be difficult to separate; however, throughout the study, there was a clear message that financial incentives attract health professionals to go to a remote region. From the perspectives of the remote health professionals, whether they positively influenced retention was contested, and difficult to distinguish from intrinsic motivation as demonstrated in the previous paragraph where the health professional who argued that money can keep you there, continued to add information about being suited to it, working in a nice community, feel safe etc., which are reasons for staying that are not related to financial incentives.

In the study it was clear that there was one group often described as a transient group of health professionals who are motivated by financial benefits and often form the short-term workforce that remote areas depend on for continuous service provision. Managers explained that there is a steady supply of these health professionals through recruitment agencies and that these agency health professionals support continuity in service delivery. Generally, agency staff commit to a time period, complete the contract, and collect the money then leave the remote region. Then the next one arrives and the pattern repeats.

In contrast, managers suggested that intrinsically motivated health professionals are more likely to stay longer, saying that intrinsically motivated health professionals usually fall into two groups: those that come to save the world and burn out quickly; and, those who find their place in the community and stay for a long time (see In Practice Box 5.2: Simon's Story). Effectively managing the latter of these groups contributes to the sustainability of remote health workforces; however, some managers found that the organisation's policies often constrained their ability to do this effectively, saying that city-centric policies can be too restrictive in remote areas making localised decision-making problematic.

[II] Reflection

Take a moment to think about your workplace.

1. How would you describe the mix of employees in your team?
2. It is sufficiently balanced or could it be improved?

Think about how you described what motivates members of your team and consider how this effects the team as a whole. If you do not think that the current balance is right, think about how it could be improved. The next time you need to recruit to your team look at this reflection and use it to inform your decision about person-fit during recruitment and selection process.

IN PRACTICE BOX 5.2: Swings and roundabouts

Simon's Story

The workforce
You can basically divide it into two camps. You have the camp that have grown up in the bush and so it's just a natural progression, it's just like putting on a comfortable pair of slippers for them, so they just fit in. But you then have the other camp, the *others*, the *outsiders*, the *urbanites*; and they can be divided further into two groups. The first group is those that go out and fall in love with it and fit in very easily and love the lifestyle of working remotely and the second group are there for other extraneous reasons and don't really like it. The second group might be there for financial gain, they might be there for career advancement, they might be there because they cannot get another job, or they might be there because they want to save everybody in the bush.

So you have got these *others*, and usually what happens is that for a nurse in remote locations, the rule of thumb is 12–18 months before they are gone. So I would get these resumes and I would interview them and of course we were desperate for staff, we needed bums on seats, boots on the ground. So anybody who applied was going to get a job and I would say to them, why do you want to come and work here? There is nothing here. And they would say, 'I have always wanted to work remote', well that is not true because you could see it in their resume they were going to get a position that they would have to wait a long time to get in a capital city. Then, as soon as they arrive, I knew that they'd be gone in 12 months because it was about career advancement. Alternatively, there would be a couple, and they would fly in and do six months work and then they would leave using the money they earned to take the next six months off, so that group had these different motives for being there.

Remote realities

Recruitment is extremely driven by pull factors because you are desperate for staff most of the time and that necessarily means that you need to accept the reality of the available workforce. The reality in remote locations is that the hospital clerk this week, may be the admin officer down at the police station next week, and the week after they will be working for the school because everyone is poaching everyone else to get the staff that they need. The other thing is that you need to, well not accept bad behaviour, but be supportive to staff, because in remote locations it is swings and roundabout.

What I mean by this is that I might have someone complaining that a nurse was not doing this or was not doing that; as a manager I think to myself well if I'm getting 80% out of someone, that is 80% more than having a vacant position and the chances of me getting someone new into the position within the next six months is low. So, I don't jump on them about the small things, I just point things out to them and work towards getting that extra 20%. So that is the reality. You have to do it because the workforce rises to meet the demands on the occasions that they are needed. For example, if there is a road accident in the middle of the night, say a bus turnover at three o'clock in the morning, there could be multiple injuries and so everybody is called out and everybody is onboard. Everybody is working under generators out on the highway for as long as is necessary. So if someone is not there on a Friday afternoon, that's fine, when they are *always* there when you need them. So, this is what I mean by it is swings and roundabouts.

Although, I am talking mainly about the second group, there is that first group, and there are a substantial amount of the workforce, rural and remote nurses, that actually love it and are fully satisfied and it is what they want to do because there are benefits you know, autonomy, extended skill bases, and flexibility. There are some wonderful benefits to remote work.

Workforce Sustainability Is Contingent on Both Intrinsic and Extrinsic Rewards

Sustainable remote health workforces are dependent on a combination of both extrinsically and intrinsically motivated health professionals which suggests a need to continue to offer financial incentives to attract and satisfactorily remunerate health professionals working in remote regions. In addition, having sufficient financial rewards needs to be complemented by intrinsic rewards. This is explained through Herzberg's motivation-hygiene theory.

Herzberg's Motivation-Hygiene Theory

Herzberg's motivation-hygiene theory proposes that,

> the primary determinants of employee satisfaction are factors intrinsic to the work that is done... These factors are called "motivators" because they are believed to be effective in motivating employees to superior effort and performance. Dissatisfaction, on the other hand, is seen as being caused by "hygiene factors" that are extrinsic to the work itself.
> (Hackman and Oldham 1976, p. 251).

In this theory 'work motivation is largely influenced by the extent to which a job is intrinsically challenging and provides opportunities for recognition and reinforcement' (Giancola 2011, p. 24). In other words, the job's context (e.g. the work itself) is far more important to employee satisfaction and motivation than hygiene factors (e.g. organisational policies) (Giancola 2011). When we consider a remote workplace, the hygiene factors must be sufficient to minimise dissatisfaction; hence, employees must feel fairly compensated for the work that they are doing, and they must have access to the resources that they require to complete this work. In remote regions, this equates to having satisfactory accommodation, remuneration commensurate with the work effort required and sufficient financial rewards to feel recognised and rewarded. As depicted in Fig. 5.1, once these hygiene needs are met; there is an opportunity for employees to experience the aspects of job satisfaction above work contentment the aspects of job satisfaction that will improve retention. However, job satisfaction cannot be improved through motivation alone if the employee's satisfaction is below the level of contentment. This explains why employees who love their job, leave. It is difficult for managers to improve retention where employees are dissatisfied with their work conditions, as one manager explained, that 'whilst we all want to be intrinsically motivated' if the accommodation was poor and 'if people weren't sure if they were going to be assaulted in the night... [it] doesn't matter what the manager does, you not going to keep them there' (Onnis 2016, p. 141).

Job satisfaction is generally associated with retention (Knights and Kennedy 2005; Ko and Hur 2014). Therefore, job satisfaction matters to health managers 'because it is

Fig. 5.1 Explaining the stages of Herzberg's motivation-hygiene theory

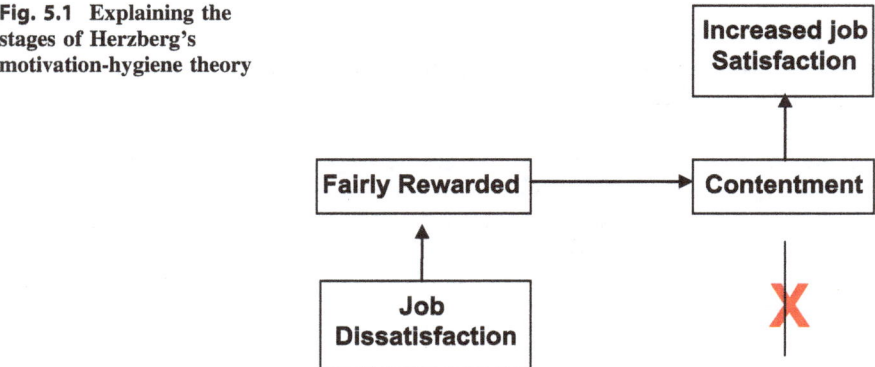

an important factor in predicting system stability (reduced turnover) and worker motivation' (Hagopian et al. 2009, p. 872). The main correlates of job satisfaction reported in the literature included: autonomy, rewards, relationships (peers and supervisors), stress, fairness, and remuneration (Penz et al. 2008). However, financial rewards may not influence retention unless the reward is at a level that required an employee to have a reduction in remuneration if they left (Wright and Kehoe 2008). As such, high remuneration and incentives can contribute to workforce instability in unintended ways, as one Remote Health Manager explained,

> 'I see a lot of people going out there, they earn pretty good money, compared to what you would if you were in metro. So you go out, you don't have all of the expenses so even though the pay is a little bit better, I think the savings potential is quite high because you don't have housing, you don't have electricity bills there's all that sort of stuff all those added savings per se and there's nothing to spend your money on really anyway. So people get used to a level of income and the struggle there is once you've had enough of working remote you move back home or back to the city which is fine, and that should be the case, but you go back and you almost halve your pay and people build their lives around their income so that becomes a real barrier and so their only way out then is to start looking for other options which is often management, because there is a higher level of remuneration, so... [it] causes people to be trapped in a remote area for a lot longer than they probably should.'

As this manager explained part of the challenge of turnover, is that often the people who need to leave do not leave and this can create further challenges for managers who can end up losing good health professionals in the process. As another remote manager explained the HRM processes can at times impede these exits, saying,

> 'I had a bad recruitment that I ended up having to performance manage, I lost good staff so that was one of my biggest learning curves of what I had done wrong. He was in a senior position, he was a bum on a seat, I was desperate, I was still green, still doing things the wrong way, and that cost our whole team actually.'

Therefore, HRM policies need to support managers so that the reward systems benefits (both extrinsic and intrinsic) can translate into improved workforce sustainability.

⑪ Reflection
Take a moment to think about health professionals in your current team or past teams that may fit the description of 'stayed too long'.

1. How did you manage the situation?
2. Was the strategy effective? How could you improve your strategy?
3. Thinking about that experience, what advice would you give yourself if you face a similar situation in the future?

References

Buykx P, Humphreys J, Wakerman J, Pashen D (2010) Systematic review of effective retention incentives for health workers in rural and remote areas: towards evidence-based policy. Aust J Rural Health 18:102–109

Campbell N, McAllister L, Eley D (2012) The influence of motivation in recruitment and retention of rural and remote allied health professionals: a literature review. Rural Remote Health 12:1900

Deci EL (1972) Intrinsic motivation, extrinsic reinforcement, and inequity. J Pers Soc Psychol 22 (1):113–120

Gagné M, Deci EL (2005) Self-determination theory and work motivation. J Organ Behav 26 (4):331–362

Giancola FL (2011) Examining the job itself as a source of employee motivation. Compensation Benefits Rev 43(1):23–29

Hackman JR, Oldham GR (1976) Motivation through the design of work: test of a theory. Organ Behav Hum Perform, 16(2):250–279. https://doi.org/10.1016/0030-5073(76)90016-7

Hagopian A, Zuyderduin A, Kyobutungi N, Yumkella F (2009) Job satisfaction and morale in the ugandan health workforce. Health Aff, 28(5):863–875. https://doi.org/10.1377/hlthaff.28.5. w863

Hegney D, McCarthy A, Rogers-Clark C, Gorman D (2002) Why nurses are attracted to rural and remote practice. Aust J Rural Health 10(3):178–186

Kanungo RN, Hartwick J (1987) An alternative to the intrinsic–extrinsic dichotomy of work. J Manag 13(4):751–766

Knights JA, Kennedy BJ (2005) Psychological contract violation: impacts on job satisfaction and organizational commitment among Australian senior public servants. Appl HRM Res 10 (2):57–72

Ko J, Hur S (2014) The impacts of employee benefits, procedural justice, and managerial trustworthiness on work attitudes: integrated understanding based on social exchange theory. Public Adm Rev 74(2):176–187

Onnis L (2014) Managers are the key to workforce stability: an HRM approach towards improving retention of health professionals in remote northern Australia. In: Proceedings, 28th Australia and New Zealand Academy of Management (ANZAM) conference. UTS, Sydney

Onnis L (2016) A sustainable remote health workforce: translating HRM policy into practice. Ph. D. thesis, James Cook University

Penz K, Stewart N, Carl DA, Morgan D (2008) Predictors of job satisfaction for rural acute care registered nurses in Canada. West J Nurs Res 30(7):785–800

Pink DH (2009) Drive—the surprising truth about what motivates us. Penguin Group, New York

Ryan RM, Deci EL (2000) Intrinsic and extrinsic motivations: classic definitions and new directions. Contemp Educ Psychol 25:54–67

Santhanam R, Hunter E, Wilkinson Y, Whiteford H, McEwan A (2006) Care, community, capacity: rethinking mental health services in remote indigenous settings. Aust J Prim Health 12(2):51–56

World Health Organisation (WHO) (2010) Increasing access to health workers in remote and rural areas through improved retention. WHO Press, France. http://www.searo.who.int/nepal/ mediacentre/2010_increasing_access_to_health_workers_in_remote_and_rural_areas.pdf. Accessed 5 Sept 2017

Wright PM, Kehoe RR (2008) Human resource practices and organizational commitment: a deeper examination. Asia Pac J Hum Resour 46(1):6–20

J.Richard Hackman, Greg R. Oldham, (1976) Motivation through the design of work: test of a theory. Organizational Behavior and Human Performance 16 (2):250-279

Amy Hagopian, Anneke Zuyderduin, Naomi Kyobutungi, Fatu Yumkella, (2009) Job Satisfaction And Morale In The Ugandan Health Workforce. Health Affairs 28 (5):w863-w875

6

Relationships: Social Exchanges, Community Ties and Employee-Manager Relationships

> *We had discovered that the manager-not pay, benefits, perks, or a charismatic corporate leader-was the critical player in building a strong workplace. The manager was the key.*
> Marcus Buckingham & Curt Coffman, The Gallup Organisation.
> (Buckingham and Coffman 1999, p. 25)

Key Messages

- Healthy employee-manager relationships, as well as professional and social relationships with peers and community members support health professionals and managers working in remote regions.
- Regardless of whether relationships are formal or emerge from situational and environmental conditions it appears that supportive relationships contribute to improvements in retention.
- Sustainable remote health workforces are more likely where employee-manager relationships are fostered, where there is perceived organisational support and where health professionals have relationship-based ties to the communities.
- A supportive employee-manager relationship becomes vital for workforce sustainability, especially for those health professionals geographically separated from their regular support networks.
- The key challenge for remote managers is balancing professional and personal relationships.

Professional and Personal Isolation

Health professionals working and living in remote regions often feel both personally and professionally isolated. As a result, a supportive employee-manager relationship becomes vital for workforce sustainability, especially for those health

© Springer Nature Singapore Pte Ltd. 2019
L. Onnis, *HRM and Remote Health Workforce Sustainability*,
Management for Professionals, https://doi.org/10.1007/978-981-13-2059-0_6

professionals geographically separated from their regular support networks (Buykx et al. 2010; Greenwood and Cheers 2002). Social exchange theory proposes that when employees perceive that the organisation treats them well, the employee reciprocates by working to benefit the organisation. Social exchange theory is presented in detail in Appendix D. In remote regions, the benefits of employee-manager relationships are important because employees who consider their manager supportive are more likely to reciprocate through commitment behaviours and loyalty to their manager which often translates into improved retention (Gould-Williams and Davies 2005; Wright and Kehoe 2008; Xerri 2013).

Health professionals working and living in remote regions often feel personally isolated when separated from family and friends (Greenwood and Cheers 2002). In the study, health professionals described feelings of personal and professional isolation experienced when working and living in remote regions and the influence of community ties. One manager said, 'the only staff that stay are the staff that have bought [housing] and have family, or their partners work in the town.' Similarly, one health professional said, 'if I didn't have a partner here now I would be less likely to stay as long as I have.' In the absence of their actual family, colleagues often reduced the sense of personal isolation, and community connections provided social support networks, e.g. 'they're leaving their friends and family behind so we become in one way their extended family.' Some managers explained that in remote regions personal characteristics and talents, well beyond their role descriptions are used to create supportive work environments, citing an example about a nurse with previous work experience as a chef who strengthened team cohesion through the sharing of food.

In the study health professionals were asked to describe how they thought their manager would describe working in remote regions. Differences in perspectives emerged. Some health professionals described managers who lacked understanding of remote work environments resulting in unsupportive relationships. Whereas, others described managers and management styles that were effective in remote regions, highlighting the significance of relationships. According to health professionals the level of management support varied, both within and between organisations. Therefore, this is an area where some remote managers can improve their relational management practices.

IN PRACTICE BOX 6.1: It Demands a Different Approach

Teresa's story

There is a psychology to managing staff, and I guess I've been able to work with the allied health disciplines long enough to know what their practice philosophy is but that warrants a different approach for each discipline just about. Some of the disciplines are very practically orientated, whereas others are very solutions focused. It demands a different approach to managing staff from a different discipline.

In my experience, managing staff in each discipline is different and I think it's really about getting to know the person and I guess in management, wherever you are, its focusing on the strengths and enabling somebody to feel that they are contributing, that they have ownership, that they've got accountability and of course are a part of the bigger team. I think that teamwork in remote services is really, really important.

Peer Support

The relationships that health professionals have with their peers can be both personally and professional supportive. Professional networks and mentoring relationships help to reduce the sense of isolation. Peer support can be formed through professional networks, teamwork and mentoring. In the study, health professionals discussed professional networks, with one allied health professional saying, 'I joined my association because that's essentially, for me it's the only contact I have with other physios.' For remote health professionals, regardless of whether they are working in multi-disciplinary teams or working in isolation as sole practitioners; professional associations can connect them to their profession to maintain their currency and to connect them to a supportive peer network.

Social support and embeddedness are considered key components for improving retention. Therefore, a workforce retention framework should include social support (Buykx et al. 2010). Some health professionals described the benefits and limitations of supportive teams, with one remote manager explaining, 'we are a quite tight team because there are *not that many of us* but then there is also *not that many of us* so I do feel more professionally isolated' (Onnis 2016). Mentoring also provides professional support and contributes to minimising the challenges associated with retention of remote health professionals (Bourke et al. 2014; Thompson 2011). The study found that community connections provide social support networks for remote health professionals. In fact, social support including the work-home interface may affect an employee's decision to leave remote regions (van der Heijden et al. 2009). Therefore, peer support, social support and embeddedness are considered key components for improving retention.

⑪ Reflection
Take some time to reflect on peer networks in your remote region. List as many peer networks as you can for your remote region.

1. How many of these peer networks have you connected with since commencing work in a remote region?
2. As a remote manager, do you continue to engage with these networks?
3. Do you need a different type of support network for your role as a manager than you did for your role as a remote health professional?
4. What is your strategy to ensure that you remain supported and engaged with peer networks relevant to your professional and personal interests?
5. How do you support your peers, and how do they support you as a remote manager?
6. Would you benefit from participating in a mentoring program?

List any actions that you could take to improve you network, and to ensure that you feel more supported in your remote manager role by your peers. Then, develop a strategy to improve your access to peer support.

☝ Resources

Take some time to explore online and see how many additional peer networks you can find. There may be Facebook groups, virtual meeting rooms, webinars, professional networks in your wider geographical region or international groups of like-minded remote health managers. Consider participating in these online networks too. Some professional organisations that offer networking opportunities and resources include:

Mt Isa Centre for Rural and Remote Health (MICRRH)
MICRRH has a network of eleven centres that provides education in Mount Isa and online using a diverse range of ICT resources. https://www.jcu.edu.au/mount-isa-centre-for-rural-and-remote-health.
Australasian College of Health Service Management (ACHSM)
The ACHSM is the peak professional body for health managers in Australasia. It brings together health leaders to network, learn and share ideas. https://www.achsm.org.au.
Australian Human Resources Institute
This Australian Human Resources Institute (AHRI), the peak body for HR professionals in Australia also offers a membership level to managers. This Australian Human Resources Institute regular hold education, training and network events across Australia, including more than 30 regional networks. https://www.ahri.com.au/conferences-and-networking/networks-and-forums/regional-and-specialist.
The Canadian Healthcare Network
The Canadian Healthcare Network aims to be the online destination for healthcare professionals to grow professionally. http://www.canadianhealthcarenetwork.ca/.

⬛ Manager's Toolkit

Explore the websites of the professional organisations listed above, as well as any others that you identify. Add any resources that think will be useful for a remote manager role to your Manager's Toolkit.

Contemporary Employment Relationships

Globally, technological and social change has had a profound impact on employment relationships, and while some aspects of human services have not significantly changed, the increased emphasis on governance and compliance systems together with technological advances and increased mobility have changed the employment landscape for remote health services. Some of the repercussions of these 21st century employment relationships are reduced organisational citizenship behaviour (OCB) and organisational commitment. OCB, discussed previously in this book, describes the behaviour displayed by employees when they work at a level beyond the scope of their role for the benefit of the organisation, for example, representing the health service in the community in their personal time without expecting compensation, speaking positively about the health service so that a good image is portrayed to the community. OCB and organisational commitment may be less evident in remote workforces, particularly where temporary employees and agency staff have reduced ties to the employer. Consequently, they are less motivated to be committed to the health service as the benefits from continued employment are absent (Coyle-Shapiro and Kessler 2000; Highhouse et al. 2007). Despite reduced organisational commitment, a focus on relationships means that remote managers have an opportunity to influence retention through supportive professional relationships.

⬛ Reflection

Revisit your reflection from chapter two about organisational citizenship behaviours and add your thoughts about how the professional relationship that you have with individual team members may influence their OCB. Make a list of the relationships that are supporting OCB. Then make a list of relationships that could be improved and any actions that you could take to improve the relationships.

IN PRACTICE BOX 6.2: It Comes Back to Bite You
Jenny's Story
I believe that the managers that don't survive are the managers who cannot manage the human characteristics as opposed to the corporate and governance issues. We have a lot of systems about what to do in an emergency, what to do in a cyclone, because you can see what needs to be done in a disaster. Most of the managers that do not last long, cannot manage the human

characteristics, and by human characteristics, I mean managing difficult people or managing personalities.

One Simple Rule
In managing difficult, complex human behaviours I have always had a very simple rule—be firm, fair, and friendly. It's really that simple. The manipulative behaviours that I have seen go on over the years, and still go on today, have been very destructive and caused significant team dysfunction. They are about survival for the individual; it is about them working out how to survive in the remote environment. Individual's feel that they best survive by doing things the way that they know and feel safe doing, and that becomes their survival. It is about how you approach them as a manager that is important to get good outcomes. It is about being consistent.

Personal and Professional Lives
So when we consider managing individuals and teams, it is about how to make sure that the people that we're working with are comfortable and confident with each other at work. A manager can build up a false impression of teamwork and that is lovely, but if it is not real, no-one is getting the support that they need from within the team. Some people like to operate with clear distinctions—this is my work and this is my private life, and the two remain separate. In rural and remote environments you may recognise this in the form of interactions out of work and in the community, such as, seeing someone in the supermarket and having them just dismiss your eye contact and smile, as if they do not know you, when you were in emergency with them just last week and you were relying on each other's skills and capability in a very intense situation. The boundaries of professional and private lives can be difficult for managers when unhealthy working relationship arise from these human characteristics.

For health professionals, providing health services absorbs quite a lot of your empathetic values, beliefs and energy. So, when they have given so much to their patients, they have little left to participate in a team building activities. For example, a health professional may not like to spend time with the people that they work with, which doesn't mean that they don't care about the patients and community, it may just mean that they have not got enough left for team care. In my experience, there is value in debriefing after you've had a significant event between two professional staff within your team who are not going to get on, and you have to make a decision as a manager about the process that you can use. Often it comes back to agreeing to disagree about the issue, and agree to act professionally in the workplace.

Support for Managers
For a manager to be consistent they need to have their own support system so they can have their bad days, and still have this approach in the workplace.

They need to have really solid support systems. I think that the use of mentoring and coaching is not well adopted in remote environments; the value of these types of professional support is not well considered or well-adopted.

The manager may need assistance and independent mediators can help to bring people's beliefs and values to the forefront for each other to understand. Once there is mutual understanding of each other's value and beliefs, you move down the road of being mutually respectful. It can be difficult for some managers because if there's nothing intrinsically wrong with the other parties beliefs and values, you are in no position to say that they are wrong. Some managers often struggle with needing to be right or wanting to change behaviour that is outside someone else's value and belief system. That is what I mean by complex and challenging human characteristics, that is what brings people down in management, it is managing the human characteristics.

Advice for Managers

There is no amount of expertise that allows you to resolve workplace issues arising from these human characteristics alone. As a manager it has always taken more than me to achieve an outcome, and that is a really important thing for any manager in rural and remote environments to remember, it's more than *you* that actually gets the outcome achieved. The care factor for health professionals is sometimes misplaced because as human beings we are always vulnerable to self-survival. To give good quality care according to need does take its toll in remote practice, to survive in remote practice requires a care factor for self as well as relationships that foster self-care. There have been staff I have had to dismiss and staff that I have had to encourage to move on over the years. And there's a big difference. Dismissal is an action that is significantly affecting people's lives. Encouraging people to move on is about making a positive impact on their life because they are not going to survive in this work environment for much longer. Good outcomes are reliant on more than *you* when you are a manager, and you better believe it because when you act alone, it always comes back and bites you!

Employee-Manager Relationship

The employee-manager relationship underpins the extent to which high commitment (or 'soft') HRM practices are implemented in the workplace (Gould-Williams 2004). Social exchange theory proposes that high commitment HRM practices shape behaviours and attitudes developing psychological links between employees and the organisation (Gould-Williams and Davies 2005). It is these links that consolidate the employee-manager relationship. In turn, a health professional's feelings toward their manager influence turnover intention (Maertz and Boyar 2012). One of the key challenges for remote managers is balancing professional and personal relationships. In the study, one remote manager explained that, 'as a manager, maintaining friendships vs collegial relationships is often very tough' and another remote manager said, 'we should have mentors... you can't bounce it off your staff' (Onnis 2016, p. 182). It is important for managers to have support from outside their team.

When considering the employee-manager relationship in remote regions the manager is responsible for managing the physical resources and meeting the community's healthcare needs, with the available workforce. In some remote regions, high turnover together with the range of health service providers often means that the remote manager may need to manage the provision of healthcare through employees from their organisation as well as employees from other organisations that provide visiting services to the community. These services are provided by health professionals that they did not employ and sometimes they are health professionals that they may not have met before that day. Therefore, they are working together despite there being no formal reporting relationships and the health professional's clinical competence being assumed but unknown to the remote manager. Thus, managing health professionals in remote regions is challenging and managers are often ill-prepared.

This lack of preparation is in no way the fault of the remote manager; it is the result of a hierarchical system where people generally rise to their level of incompetence (Taylor et al. 2010). This phenomenon, often called the 'Peter Principle', describes the circumstances where someone is promoted for being good at their job, to a position where they are unable to perform satisfactorily (Fairburn and Malcomson 2001; Pluchino et al. 2010). This could be due to the differences in the competencies required for management roles and those required in clinical roles. Regardless of whether managers consider themselves clinicians with management responsibilities, or managers with clinical responsibilities, the management philosophy that emphasises commercial practices is at odds with the philosophy of caring professions (Bolton 2003). Nowhere is this more evident than in remote community health services, where clinician-managers use management approaches reflective of clinical practices. In other words, the perceived problem (the team member) is diagnosed and an appropriate treatment implemented (Allan and Ball 2008). This is a good way to treat patients; however, it is not an ideal way to *treat* your team. Management practices should focus more on capacity building and competence of individual health professionals and the team as a whole. Hence, positive professional relationships rather than diagnosing the person's problem is a better longterm solution for team cohesion. These management skills are developed on the job, over time. Therefore, management programs, career planning and mentoring, are beneficial for developing management capability (Birks et al. 2010; Hegney et al. 2002; Thompson 2011).

⑪ Reflection

Revisit your reflection in chapter four about the skills needed for your position and how you developed them. Look at the skills on your list related to professional relationships and reflect on how you developed these skills.

1. Did you learn them in a training program or from on-the-job experiences?
2. Do you have relationship building skills that are not on the list that should be added?

💼 Manager's Toolkit

We can always learn new skills so take the time to observe others around you. If you see any actions or hear any words that you think are a great way to build relationships with others make a note of them. Over time develop this list and then put it in your Manager's Toolkit so that you can draw on it when needed. In the midst of the drama it is not always possible to think clearly, so pulling this list from your Manager's Toolkit in these challenging situations will provide you with some options that may not come to mind in the heat of the moment or when under pressure.

The Manager-Employee Relationship Influences Organisational Commitment

Social exchange theory examines the employment relationship using perceived organisational support (POS) to describe the relationship between the employee and the organisation, and leader-member exchange (LMX) to describe the relationship between the employee and their manager (Xerri 2013). Social exchange theory is described in detail in Appendix D. In brief, LMX emphasises the quality of the employee-manager relationship and is based on the degree of emotional support, and an exchange of valued resources between the employee and their manager (Ko and Hur 2014). These relationships can be influenced by management styles. Where a professional relationship exists, LMX may be linked to areas of workforce stability. This was evident in the study through comments such as, 'we had huge turnover, it was like a spinning door, that's stopped, we've got people lined up to work there now... it was quite autocratic but it was supportive' (Onnis 2016, p. 182). Therefore, management styles that influence the manner in which managers build relationships as well as the way they implement HRM policies are associated with effective localised management practices.

During the study, health professionals were asked to describe how they thought their manager would describe working in remote regions. Through this question differences in the employee-manager relationships emerged. While some health professionals described unsupportive relationships, saying, 'from their expectations of us, they had no idea', others highlighted the significance of relationships, saying, 'it just helps to have someone who knows where the communities are, what the troubles are, travelling between, how people get there, roads cut off, things like that, someone that understands the little things that make you day a lot harder' (Onnis 2017a, p. 15). Therefore, management practices that reinforce a shared understanding of the remote working environment are significant.

IN PRACTICE BOX 6.3: It Comes Down to Leadership

Peta's Story

From my personal experience as a manager, I think that it comes down to leadership. If the team feel valued and appreciated and that they are making a difference in implementing a shared vision, and I mean a real shared vision not just one that is pasted on the wall, it actually draws everyone together and they support each other, it increases peer support and it makes people feel that what they are doing is valuable and it's that intangible reward that makes people not want to leave.

This was critical in my experience as manager in a remote Aboriginal community where, as a result, the turnover rate was reduced from 200% a year to having one staff member leave in four years. That reduction in turnover was all due to leadership. I have used a similar approach in this remote community and have experienced the same thing and my team is now comprised of nearly all permanent staff. I think it is about creating a team, supporting a team and coming up with a vision that the whole team shares not just one that is imposed on them by the chief executive. It's not easy and it's not an activity that is particularly valued by the executive because they think the recruitment and retention is dealt with by financial incentives and holidays and salary sacrifice and all that stuff. Intangibles you can't report on so they are not interested in them.

A Shared Understanding

Trust is a critical factor in social exchanges and facilitates the development of social exchange relationships (Gould-Williams and Davies 2005). Researchers have analysed HR concepts such as employee engagement, job satisfaction, and organisational commitment finding that trust in the immediate manager is significant for an organisation's culture (Chalofsky 2010). Further, there is greater satisfaction and commitment when an employee's values are aligned with those of their manager (Rosete 2006). This reinforces the notion that the quality of the employee-manager relationship influences job satisfaction, commitment and turnover (Xerri 2013). Therefore, trusting, supportive relationships are essential for remote health workforce sustainability.

Employment relationships form where managers and employees are joined together by mutual dependence (Bartram 2011). In the study, health professionals felt that unless a manager has lived and worked in a remote region, they do not *really* understand what it is like to work remote, implying that without this understanding they cannot effectively manage remote health professionals. In the study, frustrations about unrealistic expectations based on a lack of awareness about the work environment were evident. One health professional said that 'the city based staff have no understanding of conditions, resources and remote areas' and another said that they were 'tired of having to follow directives from Metro/Regional based managers' (Onnis 2016, p. 194). While statements like these were directed at city-based managers, others reported a lack of awareness from remote managers with one health professional's frustration evident when asked in an interview for five words that their manager would use to describe working in a remote region. In response the health professional said 'I just have to say they did not know... That's four words; make it capitals THEY HAD NO IDEA!' and another said that the urban-based managers were 'often out of touch with their expectations' (Onnis 2017a, p. 5). One health professional offered solutions suggesting that there is a need train managers in distance management saying, 'just because they are able to manage a team face-to-face doesn't mean they have any capability of managing a team from a distance. My experience being out bush was that you only ever heard from a manager when they wanted to yell at you.' These comments highlight the role of the employee-manager relationship in localised management practices.

⑪ Reflection
Take a moment to answer the questions posed in the study.

1. What are the five words that your manager would use to describe working in a remote region?
2. What are the five words that you would use to describe working in a remote region?
3. What five words do you think that a member of your team would use to describe working in a remote region?

Look at the three sets of five words that you have written.

1. Are they different?
2. Which words were the same?
3. Reflect on what this means, and how you could improve your understanding/or assist others to understand what it is like to work in a remote regions.

You may like to discuss this with you team as a way of building a shared understanding of what it is like to work in a remote regions from your different perspectives. Find the advertisement that you wrote in Chapter four. Revisit the description and think about how well it captures your understanding of what it is like to work in a remote region.

 **Resources**

What Can We Learn About Improving Workforce Retention from Five Words?

You can see the lists of words that other remote managers and remote health professionals used to describe working in remote regions in this conference paper which was presented at the 14th National Rural Health Conference, http://www.ruralhealth.org.au/14nrhc/program/concurrent-speakers.

IN PRACTICE BOX 6.4: We Listen to Staff, We Enable Them to Have a Voice

Sarah's Story

We enable staff to have a voice; we have got a communication plan so that communication is kept very open. Staff meet with their manager monthly and certainly there is a focus on them meeting their KPIs but there is also a focus on asking about them personally, e.g. How are you going? Are you OK? What do you need to deliver your job? How are you managing? Are you happy? We bend over backwards for our staff, for example, if they needed to have time out, we have a workplace reward policy which means that we support staff financially if they want to join a gym or they want to go to a yoga class or if they want to do something else around being healthy and looking after themselves.

Part of the performance review is making sure that people have a good work-life balance and we work with them to find that balance. It is not all left to them, because usually they have got their head down and part of management is being able to step in and talk to other managers at clinics and try to work through a better way to do things. I think also, just being very clear about the directives of the organisation and how we are going to go there. So we are busy at the moment we've just got a new strategic plan, now operationalising that and being realistic about our expectations of staff. And again letting them have input and ownership into what we plan to do. We're not a big organisation; we haven't got that whole multi-tier of hierarchy so we can be personal with our staff. I know every staff member, I know their family, I know where they've come from and I have a good sense of their strengths and it would not be unlike me to ring and say, 'Hey, I heard about this, just thought I would let you know'. So I take a personal interest in the affairs of my staff and so do the other managers.

References

Allan J, Ball P (2008) Developing a competitive advantage: considerations from Australia for the recruitment and retention of rural and remote primary health workers. Aust J Primary Health 14 (1):106–112

Bartram T (2011) Employee management systems and organizational contexts: a population ecology approach. Manage Res Rev 34(6):663–677

Birks M, Mills J, Francis K, Coyle M, Davis J, Jones J (2010) Models of health service delivery in remote or isolated areas of Queensland: a multiple case study. Aust J Adv Nursing 28(1):25–34

Bolton S (2003) Multiple roles? Nurses as managers in the NHS. Int J Public Sector Manage 16 (2):122–130

Bourke L, Waite C, Wright J (2014) Mentoring as a retention strategy to sustain the rural and remote health workforce: a rural mentoring model. Aust J Rural Health 22(1):2–7

Buckingham M, Coffman C (1999) First, break all the rules. In: What the worlds greatest managers do differently. Simon and Schuster, London, UK

Buykx P, Humphreys J, Wakerman J, Pashen D (2010) Systematic review of effective retention incentives for health workers in rural and remote areas: towards evidence-based policy. Aust J Rural Health 18:102–109

Chalofsky NE (2010) Meaningful workplaces: integrating the individual and the organization. Jossey-Bass, San Fransisco, United States

Coyle-Shapiro J, Kessler I (2000) Consequences of the psychological contract for the employment relationship: a large scale survey. J Manage Stud 37(7):903–930

Fairburn JA, Malcomson JM (2001) Performance, promotion, and the peter principle. Rev Economic Stud 68(1):45–66

Gould-Williams J (2004) The effects of 'high commitment' HRM practices on employee attitude: the views of public sector workers. Public Adm 82(1):63–81

Gould-Williams J, Davies F (2005) Using social exchange theory to predict the effects of HRM practice on employee outcomes. Public Manage Rev 7(1):1–24

Greenwood G, Cheers B (2002) Doctors and nurses in outback Australia: living with bush initiatives. Rural Remote Health 2:98

Hegney D, McCarthy A, Rogers-Clark C, Gorman D (2002) Why nurses are Resigning from rural and remote Queensland health facilities. Collegian 9(2):33–39

Highhouse S, Thornbury EE, Little IS (2007) Social-identity functions of attraction to organizations. Organizational Behav Human Decis Process 103(1):134–146

Ko J, Hur S (2014) The impacts of employee benefits, procedural justice, and managerial trustworthiness on work attitudes: integrated understanding based on social exchange theory. Public Adm Rev 74(2):176–187

Maertz CP, Boyar SL (2012) Theory-driven development of a comprehensive turnover-attachment motive survey. Human Res Manage 51(1):71–98

Onnis L (2016) A Sustainable Remote Health Workforce: Translating HRM Policy into Practice. Ph.D. Thesis. James Cook University

Onnis L (2017a) Human resource management policy choices, management practices and health workforce sustainability: remote Australian perspectives. Asia Pacific J Human Res (early online)

Onnis L (2017b) What can we learn about improving workforce retention from five words? In: Coleman L (ed) Proceedings of the 14th national rural health conference, Cairns, Queensland, 26–29 March 2017. National Rural Health Alliance, Canberra

Pluchino A, Rapisarda A, Garofalo A (2010) The peter principle revisited: a computational study. Phys A 389(3):467–472

Rosete D (2006) The impact of organisational values and performance management congruency on satisfaction and commitment. Asia Pacific J Human Res 44(1):7–24

Taylor R, Blake B, Claudio F (2010) An analysis of the opinions and experiences of Australians involved in health aspects of disaster response overseas to enhance effectiveness of humanitarian assistance. University of Queensland, Queensland, Australia

Thompson P (2011) The trouble with HRM. Human Res Manage J 21(4):355–367

van der Heijden BIJM, van Dam K, Hasselhorn HM (2009) Intention to leave nursing: the importance of interpersonal work context, work-home interference, and job satisfaction beyond the effect of occupational commitment. Career Develop Int 14(7):616–635

Wright PM, Kehoe RR (2008) Human resource practices and organizational commitment: a deeper examination. Asia Pacific J Hum Res 46(1):6–20

Xerri M (2013) Workplace relationships and the innovative behaviour of nursing employees: a social exchange perspective. Asia Pacific J Hum Res 51(1):103–123

Resourcing: Access, Availability, and Localisation

7

I took this job to learn how to lead in difficult situations so I'll keep reminding myself that the idea is to be comfortable with ambiguity.

Rachael Robertson, Antarctic Expedition Leader.
(Robertson 2014, p. 264)

Key Messages

- The availability of resources has implications on performance and workforce stability.
- The most frequently discussed resourcing concern was staff accommodation.
- The nature of the work creates additional challenges; therefore, in remote regions personal support often extends beyond the traditional work day.
- While many of the physical hazards are associated with the remote geography, some of the mental health and general wellbeing issues are avoidable, and at a minimum could be reduced through improved management practices, work systems, and access to resources.
- Management practices that strive to localise policies without compromising the integrity of health services are better positioned to create sustainable workforces.

Implications of Poor Resourcing

'In remote communities, health professionals need to have private use of the vehicle when they're in a job, including the locums. If you don't have access to a vehicle, you are stuck in a house after hours. I didn't want to be stuck inside a house for the whole time that I am

© Springer Nature Singapore Pte Ltd. 2019
L. Onnis, *HRM and Remote Health Workforce Sustainability*,
Management for Professionals, https://doi.org/10.1007/978-981-13-2059-0_7

there, except for the hours I am in the clinic. I want to find out about how the community operates and the geography and the birdlife and flora and fauna of this area. I have to say the jobs when I had access to a vehicle were more enjoyable placements.'

Remote Health Professional

Health professionals are attracted to work in remote regions for a variety of reasons discussed earlier in this book; however, this attraction does not always prepare them for the challenges they experience working in regions where there are limited resources (Devine 2006; Opie et al. 2011; Wakerman et al. 2009). Limited resources increases the cost of living, limits access to leave, and on-call which are often intensified by workforce shortages and inadequate financial and staffing resources. As a result, managers need to be flexible, organised, creative, have good time management, prioritisation, and general management skills if they are to be effective with limited resources.

In general, health professionals report dissatisfaction with the lack of clinic facilities and resources, restricted access to transport, poor co-ordination of visiting services and overnight accommodation (Battye and McTaggart 2003; Birks et al. 2010; Devine 2006; O'Toole and Schoo 2010; Santhanam et al. 2006). In the study, health professionals elaborated on their frustrations with poor resourcing, such as, inadequate infrastructure and technology, restricted access to vehicles, and outdated equipment, saying that they 'didn't realise until moving to a remote region how limited resources can be for work', and one manager commented on infrastructure, saying, 'internet speeds are slow, too slow, or non-existent.' Another health professional explained that restricted access to vehicles extends beyond reducing their work-related efficiency and into their personal space by reducing their job satisfaction.

While limited resources has implications of the effectiveness of the health services provided, they also have health implications for individuals, the extent to which this impact is observed differs depending on individual personal coping strategies, health regimes and personal resilience. As one remote manager explained, 'We're isolated, we don't have the normal commodities around us so you need to be good with your own time and entertaining yourself and you need to be tough to deal with some of the challenges of remote service delivery.' The most frequent health implications discussed in the literature included stress, distress, feeling exhausted and overwhelmed, tired, fatigued and burnout (Birks et al. 2010; Carey 2013; Devine 2006; Gardiner et al. 2005; Kruger and Tennant 2005; Lenthall et al. 2011; Opie et al. 2011). Overall, the organisation needs to ensure that there are adequate resources. This includes having the available resources for orientation and supporting health professionals in their early days in the remote workplace. One remote manager explained that her approach to recruiting new staff depends on having adequate staffing to support them:

'We would only put on someone new, who wasn't as experienced, if the rest of the team were experienced and could support them. If there's no-one to support them then we'd think twice about taking that person on. We want it to be successful for them and for us, so it's really an individual decision. But if we take someone on, even if they are experienced, they get a thorough orientation. So we try to send clinicians out in groups, so it may be that the podiatrist, the dietician and the diabetes educator go out together. The team are such that they will look out for one another and do introductions, and everything. I am totally confident that my staff would bend over backwards to make that new person feel welcomed, and not in too deep so that they were really feeling quite anxious about it all.' Remote Manager.

In contrast, other health professionals, explained the negative impact of limited staff resources, with one saying that 'there's only five nurses who can do oncall so they are pretty much oncall all the time' (Onnis 2016, p. 160, 2017b). Managers reported similar challenges saying, 'working during the day and doing oncall at night that takes a toll on people' and 'having access to leave and being able to use it are two completely different things' (Onnis 2017b, p. 17). Moreover, limited funding for backfill and delays in recruiting increase pressure on existing staff support visiting services and covering vacancies (Battye and McTaggart 2003).

⑪ Reflection
In your reflection journal, consider how resourcing challenges influence your workplace and the health services that you offer.

1. What are the implications of poor resourcing for your team?
2. How does access to adequate resources shape you management practice?
3. What strategies do you use to compensate for the resourcing challenges?
4. How do the resourcing challenges affect your team and the health services that you can provide for your community?

Accommodation

'[T]he accommodation is interesting because you'll have people who are quite fine about where they stay, they don't mind; and then others, they want internet connection, they want this, they want that' HR Manager.

The most frequently discussed infrastructure concern in the study was staff accommodation. Most health professionals and managers described accommodation as being expensive, poor quality, inequitably distributed and often did not meet expectations. One health professional described the nursing accommodation saying, 'the quality of the accommodation was so poor… in one room flats, like bedsits out the back, in the compound of the clinic, in the backyard behind the clinic' (Onnis 2017a, p. 183). In contrast, a few had satisfactory accommodation, which was free or subsidised by their employer, with one doctor described their accommodation, saying 'we've got a great house … three bedroom, air con in every bedroom, we've got a water view … accommodation was not an issue' (Onnis 2017b, p. 16).

However, the study, found that accommodation was often an issue. Particularly, when there were unrealistic expectations. Even when health professionals provide assurances that they are familiar with remote work environments, they can have unrealistic expectations as explained by one remote manager:

'well it was often accommodation that was not what they were expecting and accommodation for staff in remote areas is very hard. It's very difficult, it's a lot of money, it's
improving but, so often the accommodation wasn't up to standard or the clinic wasn't what
they expected. Or the equipment wasn't what they expected despite the fact that you tried
your best in the interview to explain all these and they assured you that they worked remote
[previously]... But really when push came to shove and they were on call that night they
couldn't cope and they'd be on the phone to you saying I can't do this... and then when you
phone back you don't get any response so you're up all night worried about them and head
off the next day... you just have to remove people from the situation there because it's not
doing them any good and it certainly not doing you any good or the service any good. But it
is difficult... some staff threaten to leave because they don't get what they expect they are
entitled to, as far as accommodation or clinics go, but I think you have to try going out of
your way to make people happier in the role if they're discontented but at the same time
you have to tow the line and say this is being unrealistic.'

The analysis of the recruitment advertising found references to the type of
accommodation which restricted the type of applicants that would be attracted to
live and work in the community, e.g. 'Most of our clients offer free single
accommodation (bringing a partner is not recommended)' and 'This accommodation is for employees ONLY and therefore cannot accommodate partners, children
or pets.'

There is no easy solution to the accommodation challenges which will indu-
bitably continue to be a challenge for remote managers for some time yet. However,
the story from the remote manager and the HR Manager in this section of the
chapter highlights the role that remote managers must play in distinguishing
between unrealistic expectations and the normal accommodation challenges. The
previous narrative about realistic recruitment practices becomes important in this
discussion too. If the recruitment advertising contributes to the creation of unrealistic expectations and the formation of psychological contracts about the type of
accommodation the employer is obligated to provide; it is not surprising that the
remote manager finds themselves dealing with a new recruit who is unhappy with
their accommodation. Therefore, more attention to the types of expectations created
during the recruitment process can be beneficial in preventing some of the challenges that arise with new employees in regards to their preconceived expectations
for accommodation. However, as the remote manager's story demonstrates,
sometimes no matter how much you try to prepare someone, they may still arrive
with unrealistic expectations and a sense of entitlement that is not a good fit for your
workplace. As such, the remote manager will need to manage the situation towards
the best outcome for the community, the health service and the health professional.

[I] Reflection

Consider the accommodation situation in your remote workplace.

1. What are challenges that you experience with accommodation?
2. Is it equitably distributed or do some health professions (e.g. doctors) have
   different entitlements than other remote health professions (e.g. nurses, local
   health workers)?

3. What strategies do you use to minimise unrealistic expectations about accommodation?
4. Are there any further actions that you can take to more equitably distribute accommodation?

Workplace Health, Safety and Wellbeing

Perceived Organisational Support (POS), which was introduced in previous chapters and is explained in detail in Appendix D, is a form of social exchange between an employee and their employer (Wright and Kehoe 2008). When organisations treat employees well, employees reciprocate by working to benefit the organisation (Brunetto et al. 2013). POS forms through supportive management practices, such as providing for personal safety (e.g. when organisations provide safe workplaces to meet occupational health and safety requirements). However, through supportive employee-manager relationships there is an opportunity to improve workforce sustainability by providing for both the physical and mental health of employees. In remote regions, safety concerns, particularly fatigue (associated with travel and excessive oncall work at night), workplace bullying, burnout, and violence are management priorities. One manager explained the hazards, saying, 'there are a lot more safety aspects that we need to consider up here. We've got people driving, by themselves ... They could be stuck somewhere out by themselves in the middle of nowhere' (Onnis 2016, p. 183). Regardless of whether you are working in outback Australia, Canada, China, India or the US, the nature of remote work increases the time spent travelling (roads or air) over remote terrain which subsequently increases the risk of work-related injuries (Newhook et al. 2011).

Management responsibilities extend to the provision of a safe work environment; however, often the required skills take time to develop and with experience often comes the wisdom to know how and what aspects of personal behaviour to manage to ensure a safe workplace for everyone. For some remote managers, their clinical skills are transferable, such as, the communication and interpersonal skills developed to de-escalate and maintain a safe environment with clients (e.g. not raising your voice, using clear and simple language and remaining calm) (Sim et al. 2011). However, for remote managers, additional management skills are needed to negotiate the complex challenges that ensure the personal safety of remote health professionals.

In response to violent events in remote communities in Australia, there has been increased effort to investigate and identify how safety and security can be improved for the remote health workforce. The *Working Safe Project* aimed to provide managers and health professionals working in rural and remote Australia with information, guidelines and strategies all of which can be adapted and localised to a particular remote working environment. The website (www.worksafe.com.au) contains information that is valuable for both managers and health professionals working in remote regions. In addition, the *Working Safe in Rural and Remote*

Australia Project (conducted by CRANAplus) primarily focused on safety risks and violence for the remote nursing workforce and provides further information and resources for remote managers and health professionals. The Working Safe in Rural and Remote Australia Project Report found that 'being female, in or around your own accommodation, and after hours' times are risk factors'(CRANAplus 2017, p. 6). The report recommends that organisations recognise these risk factors in staff orientation. Further, they propose that industry specific literature has focused on violence, to the detriment of other significant threats to remote health workforce safety and security. As a result, there is limited literature about successful strategies and benefits arising from the successful implementation of strategies, policies and practices aimed at improving the security and safety of the remote workforce, particularly after hours. The report highlighted the issues where further research and intervention are needed, including: 'Vehicle and travel safety; dog attack; bullying and harassment; and personal health and wellbeing' (CRANAplus 2017, p. 6). For the remote manager, this focus on safety and a focus of future research towards strategies, policies and practices that can improve safety will provide the remote manager with guidance about how best to manage these situations; and more importantly how to implement preventative strategies so that they can be avoided.

While many of the physical hazards are associated with the geography, some of the mental health and wellbeing issues are avoidable, and could be reduced through improved management practices that coordinate work systems, and access to necessary resources. In fact, McCullough et al. (2012b, pp. 7–8) proposed that a 'lack of action from management when hazards are identified by clinic staff and insufficient recognition of the risk of violence by employers were also significant hazards', suggesting that remote managers that do not take action are in themselves a risk factor. Therefore, orientation and ensuring that new staff have the resources they need, and have a realistic understanding of the environment in which they have been recruited to work is an essential component of workplace health and safety that has direct implications on workforce sustainability.

To provide remote managers with assistance to develop skills in this area, McCullough et al. (2012a, p. 329) developed a management toolbox of strategies using 'toolbox' as an analogy to acknowledge the 'multifaceted nature of occupational violence and the need for a range of strategies to reduce the risk of violence towards RANs.' They proposed that 'Reducing the risk of occupational violence might reduce staff turnover and the consequences of skill shortage in remote areas' (McCulloch et al. 2012a, pp. 332–333). In the study, health professionals often reported instances where access to information about the remote community and work systems, for new employees was lacking. This is exemplified in Anne's story (In Practice Box 4.2).

The individual wellbeing of health professionals influences the quality of care that they are physically and emotionally able to provide to their clients, as well their ability to work competently, to the best of their capacity. Therefore, the wellbeing of health professionals is a personal, professional, and an organisational responsibility. The WHO Report on retention in rural and remote regions recognises the importance of personal and professional support as key factors influencing the retention of health professionals (Hegney et al. 2003; WHO 2010).

The benefits of strong relationships between employee wellbeing and voluntary turnover are known (Page and Vella-Broderick 2009). Similarly, the effects of the absence of support in challenging circumstances, is known to lead to voluntary turnover and have negative health impacts for the individual(s) involved (Xerri 2013). Further, even though the negative effects of bullying, harassment and lateral violence are well documented (Wright and Kehoe 2008) they continue to be present in workplaces. In remote regions, inadequate management responses can place further pressure on a health service already under resourced. Therefore, management practices that do not appropriately respond to ensure the personal safety of remote health professionals can have long-term impacts for the individual health professional and their families, as well as societal wellbeing (Brunetto et al. 2013). Furthermore, events that arise from the inadequate provision of personal safety can reduce the attractiveness of remote workplaces for future health professionals.

Health professionals working in remote regions face unique pressures due to geographical, personal and professional isolation. Therefore, providing support improves their professional competence, personal wellbeing, and promotes workforce stability. Health professionals who feel adequately supported have greater capacity to positively influence their own personal wellbeing, as well as the health outcomes for remote populations. Therefore, management support is not only a significant aspect of retention for organisations, it also promotes workforce sustainability, which benefits individuals and remote populations through better access to appropriate health services.

[||] Reflection

Consider workplace health and safety in your workplace.

1. Do you know the role and responsibilities that you have a remote manager?
2. Do your management practices focus on meeting legislative requirements or do you consider the health and wellbeing aspects for you team more holistically?
3. When did you last discuss health and wellbeing with your team?
4. Do you suspect that bullying exists in your workplace? If so, what actions have you taken? What actions will you take in the future if you become aware of bullying? Are you aware of your responsibilities as a remote manager if bullying is present and you are aware of it in your workplace?
5. How is safety described in the workplace orientation to your remote workplace?
6. Does your team have access to tools to support their personal health and wellbeing while working in your remote workplace?

Take a moment to reflect on your answers and consider the actions you need to take to ensure the safety of you, your team, and the community.

 **Resources**

This section provides a summary of selection of resources about safety, health and wellbeing currently available to remote managers. There has been a lot of activity around safety for remote health professional recently, and it is most likely that there will be more activity in the future, so it is recommended you source new information, tools and resources for your workplace regularly.

Working Safe in Rural and Remote Australia
The Working Safe project aims to provide managers and health professionals working in rural and remote Australia with information, guidelines and strategies all of which can be adapted and localised to a particular working environment. http:// workingsafe.com.au/

Remote Health Workforce Safety Security Report (CRANAplus 2017).
This report is available at: https://crana.org.au/uploads/pdfs/Remote-Health-Workforce-Safety-Security-Report-January-2017-5c4e6cc07ef30b87cf919ca42084a0a0.pdf

Working Safe in Remote and Isolated Health Handbook
The handbook provides an introduction to remote health safety and security for those starting out. https://crana.org.au/media/news/2017/working-safe-in-remote-and-isolated-health-handbook

Manager's Toolkit
You may wish to add some of these resources to your Manager's Toolkit.

Working Safe in Remote Practice (Free online e–remote module #13)
This module provides an introduction to remote and isolated health safety and security issues. https://crana.org.au/education/eremote/programs/core-mandatories

Violence management toolbox
This article is a good resource developed for managers working in remote areas who wish to proactively reduce the risk of violence in their workplace:
    McCullough K., Lenthall S., Williams A., Andrew L (2012a) Reducing the risk of violence towards remote area nurses: A violence management toolbox. Australian Journal of Rural Health, 20, 329–333

Personal support
Please add your organisation's Employment Assistant Program (EAP) and any local services that you know about to the list of personal support services for your Manager's toolkit. Then, if you, or a member of your team, feel that this type of support would be beneficial, the details are available.

Australia
Beyondblue—Support for anyone feeling depressed or anxious: 1300 224 636
Bush Support Services—Support for the remote health workforce and their families: 1800 805 391
Headspace—Mental health service for young people (12–25 years): 1800 650 890
Lifeline—Support for anyone having a personal crisis: 13 11 14
Suicide Call Back Service—Support for anyone thinking about suicide: 1300 659 467

Canada
Crisis Services Canada: www.crisisservicescanada.ca
1-833-456-4566, text 45645, or residents of Quebec, 1-866-277-553.
Mental Health crisis line: https://www.crisisline.ca/ Ottawa: 613-722-6914

United Kingdom
The Samaritans operates a free to call service 24 h a day, 365 days a year: 116 123
or visit https://www.samaritans.org/how-we-can-help-you/contact-us
Apps and digital tools to help you manage: https://apps.beta.nhs.uk/?category=
Mental%20Health
Moodzone: https://www.nhs.uk/conditions/stress-anxiety-depression/

New Zealand
Mental Health Foundation of New Zealand: https://www.mentalhealth.org.nz/get-help/in-crisis/
Free call or text 1737; Lifeline: 0800 543 354 or (09) 522 2999. Free text 4357 (HELP) or
Youthline: 0800 376 633; Samaritans: 0800 726 666

Europe
Mental Health Europe: https://mhe-sme.org/

Indigenous Social and Emotional Wellbeing Programs
The Family Wellbeing Program is an Australian Aboriginal developed empowerment program that is run by Aboriginal and Torres Strait Islander people for Aboriginal and Torres Strait Islander people. This story-telling style empowerment program supports participants in developing leadership skills. https://family-wellbeing.squarespace.com

Localised Management Practices Shape Workforce Sustainability

The aspects of poor resourcing described in this chapter, reinforce the need for effective localised management practices. Chapters 4–7 explored four HRM challenge areas where management practices can shape workforce sustainability: recruitment, remuneration, relationships and resources. All four HRM areas are interrelated; therefore, it is through the development and implementation of localised HRM policies, managers strengthen relationships and form a deeper understanding of the challenges and rewards of working in remote regions.

In the study, managers explained that HRM policies need to be localised. They suggested that policies need to be contextualised when applied to the local area, as often new remote managers come in and disregard something because it is done differently in the city, without knowing why it is done that way locally. The influence of management practices and the localisation of HRM policies during implementation is important. Hence, management practices that localise HRM policies without compromising the integrity of health services are better positioned to create sustainable workforces. Throughout Chaps. 4–7, the influence of management practices has been evident. The next section of this book takes the findings from the study, and the HRM theories mentioned in the previous chapters (Appendices B–E), and discusses an HRM framework developed to assist remote managers in understanding how management practices shape workforce sustainability.

References

Battye KM, McTaggart K (2003) Development of a model for sustainable delivery of outreach allied health services to remote north-west Queensland, Australia. Rural Remote Health 3:194

Birks M, Mills J, Francis K, Coyle M, Davis J, Jones J (2010) Models of health service delivery in remote or isolated areas of Queensland: a multiple case study. Aust J Adv Nurs 28(1):25–34

Brunetto Y, Xerri M, Shriberg A, Farr-Wharton R, Shacklock K, Newman S, Dienger J (2013) The impact of workplace relationships on engagement, well-being, commitment and turnover for nurses in Australia and the USA. J Adv Nurs 69(12):2786–2799

CRANAplus (2017) Remote Health Workforce Safety and Security Report: Literature review, Consultation and Survey report. CRANAplus, Cairns

Carey TA (2013) A qualitative study of a social and emotional well-being service for a remote Indigenous Australian community: implications for access, effectiveness, and sustainability. BMC Health Serv Res 13:80

Devine S (2006) Perceptions of occupational therapists practising in rural Australia: a graduate perspective. Aust Occup Ther J 53(3):205–209

Gardiner M, Sexton R, Durbridge M, Garrard K (2005) The role of psychological well-being in retaining rural general practitioners. Aust J Rural Health 13(3):149–155

Hegney D, Plank A, Parker V (2003) Workplace violence in nursing in Queensland, Australia: A self-reported study. Int J Nurs Pract 9(4):261–268. https://doi.org/10.1046/j.1440-172X.2003.00431.x

Kruger E, Tennant M (2005) Oral health workforce in rural and remote Western Australia: practice perceptions. Aust J Rural Health 13:321–326

Lenthall S, Wakerman J, Opie T, Dunn S, MacLeod M, Dollard M, Rickard G, Knight S (2011) Nursing workforce in very remote Australia, characteristics and key issues. Aust J Rural Health 19(1):32–37

McCullough K, Lenthall S, Williams A, Andrew L (2012b) Reducing the risk of violence towards remote area nurses: A violence management toolbox. Aust J Rural Health 20:329–333

McCullough K, Williams A, Lenthall S (2012a) Voices from the bush: remote area nurses prioritise hazards that contribute to violence in their workplace. Rural Remote Health 12:1972

Newhook J, Neis B, Jackson L, Roseman S, Romanow P, Vincent C (2011) Employment-Related Mobility and the Health of Workers, Families, and Communities: The Canadian Context. Labour (Spring):121–156

Onnis L (2016) A Sustainable Remote Health Workforce: Translating HRM Policy into Practice. Ph.D. Thesis. James Cook University

Onnis L (2017a) Attracting Future Health Workforces in Geographically Remote Regions: perspectives from current remote health professionals. Asia Pacific Journal of Health Management 12(2):25–33

Onnis L (2017b) Human resource management policy choices, management practices and health workforce sustainability: remote Australian perspectives. Asia Pacific Journal of Human Resources (early online)

Opie T, Lenthall S, Wakerman J, Dollard M, MacLeod M, Knight S, Rickard G, Dunn Sandra (2011) Occupational stress in the Australian nursing workforce: a comparison between hospital based nurses and nurses working in very remote communities. Aust J Adv Nurs 28(4):36–43

O'Toole K, Schoo AM (2010) Retention policies for allied health professionals in rural areas: a survey of private practitioners. Rural Remote Health 10:1331

Page KM, Vella-Brodrick DA (2009) The 'What', 'Why' and 'How' of Employee Well-Being: A New Model. Soc Indic Res 90(3):441–458

Robertson R (2014) Leading on the edge. Wiley, Australia

Santhanam R, Hunter E, Wilkinson Y, Whiteford H, McEwan A (2006) Care, Community, Capacity: Rethinking Mental Health Services in Remote Indigenous Settings. Australian Journal of Primary Health 12(2):51–56

Sim MG, Wain T, Khong E (2011) Aggressive behaviour: Prevention and management in the general practice environment. Australian Family Physician 40(11)

Wakerman J, Humphreys JS, Wells R, Kuipers P, Jones JA, Entwistle P, Kinsman L (2009) Features of effective primary health care models in rural and remote Australia: a case-study analysis. Med J Aust 191(2):88–91

World Health Organisation (WHO) (2010) Increasing access to health workers in remote and rural areas through improved retention.WHO Press, France. http://www.searo.who.int/nepal/mediacentre/2010_increasing_access_to_health_workers_in_remote_and_rural_areas.pdf Accessed 5 September 2017

Wright PM, Kehoe RR (2008) Human resource practices and organizational commitment: A deeper examination. Asia Pacific Journal of Human Resources 46(1):6–20

Xerri M (2013) Workplace relationships and the innovative behaviour of nursing employees: a social exchange perspective. Asia Pacific Journal of Human Resources 51(1):103–123

Part III
Localising Management Practices

An Integrated HRM Framework for Remote Managers

8

The world is made in circles. And we think in straight lines.
Peter Drucker, Founder, Society for Organizational Learning.

Key Messages

- Supportive management practices are more likely to improve workforce sustainability.
- The HR outcomes that influenced remote health workforce sustainability were associated with professional isolation, empowerment and remuneration.
- Without management practices HRM policy choices do not result in the HR outcomes that influence remote health workforce sustainability.
- The *I*-HRM-SRHW framework can assist remote managers to implement localised HRM policies and improve workforce sustainability.
- Remote managers need to know the key ingredients for workforce sustainability; then determine the combination and quantity of these ingredients suitable for their particular remote context.

Reframing and Refocusing

'If you have a turnover problem, look first to your managers.'

(Buckingham and Coffman 1999, p. 27)

For many years, the attention on workforce shortages in remote regions worldwide has been on high turnover rates and the difficulties in attracting and retaining health professionals in less desirable geographical locations. Consequently, the focus has been on how to reduce turnover and how to improve

© Springer Nature Singapore Pte Ltd. 2019
L. Onnis, *HRM and Remote Health Workforce Sustainability*,
Management for Professionals, https://doi.org/10.1007/978-981-13-2059-0_8

retention. However, turnover is an outcome; it marks the end of an employment relationship. This means that managers need to reframe their focus toward shaping the employment relationship in ways that reduce the likelihood that voluntary turnover is the natural outcome. In this book it is argued that the remote manager is the person who can influence employment outcomes. Therefore, the third section of the book focuses on how localised management practices influence workforce sustainability in remote regions.

The first section of this book synthesised what was known about the challenges and rewards of working in remote regions. Revealing that some studies suggested that poor management practices and ineffective human resource management practices were associated with outcomes such as employee burnout, poor population-level health and workforce turnover (Birks et al. 2010; Brunetto et al. 2010; Lenthall et al. 2009). Then, using a strengths-based approach, the focus turned from turnover to workforce sustainability. The second section of the book focused on the HRM challenges, showing how management practices look *in practice* with examples and descriptions from health professionals working in remote regions and managers who are managing remote health professionals. Several studies reported on the impact of poor management practices (Birks et al. 2010; Lenthall et al. 2009); however, few focused on how an increase in management support could improve workforce sustainability. Hence, this section focuses on how managers can better manage remote workforces building on the findings from the study including an HRM framework developed specifically for remote managers.

Tailoring Training

The context is important for managers, and this is especially true for remote managers. When localising management practices, instinctively remote managers will be seeking training specific to remote regions. Management development and training tailored to the remote context is important; however, managers should also consider whether generic management training is transferable to their remote workplace. In her account of her experience as the Station Manager in Antarctica, Robertson (2014, p. 78) commenting on the lack of specific training for a manager working in a geographically isolated location, said, 'I thought that surely such a harsh and uncompromising environment would create unique management scenarios that aren't found in the average workplace'. Robertson (2014, p. 78) went on to say, 'Towards the end of my time in Antarctica I would reflect on the leadership and management challenges and agree. They are no different in principle, but they are vastly different in application!' Robertson, emphasised the importance of the application of management practices, and that leadership is a key aspect of effective management practices, as was pointed out by one health professional in the study, who said,

'[R]emote services have the opportunity to be more flexible and innovative than large metropolitan services but they need good leadership and the options to do things in ways that work well for their particular areas. Sometimes it just does not work trying to apply models that work well in other parts of the country.' (Onnis 2016, p. 146)

Hence, it is leadership and the application (or implementation) of HRM policies in remote regions that determine a manager's effectiveness in their remote workplace. This book contains activities and resources to assist remote managers to develop their skills in different areas of HRM. These activities are not only supporting skill development, they provide opportunities for remote managers to adapt the suggested activities to their own remote context. The items that have been added to the Manager's Toolkit have been customised by *you* to suit *your* context prior to putting them in *your* toolkit. This part of the process is important, as it is in a sense the start of the 'localisation' of your management practices. This is a practice that will be valuable throughout your management career.

Supportive Management Practices

'While some HR policies may impact on employees directly, most rely on line manager action or support, and the quality of the relationship between employees and their immediate line managers is liable, too, to influence perceptions not only of HR practices but of work climate, either positively or negatively.' (Purcell and Hutchinson 2007, p. 5)

The available evidence suggests that a coordinated and well-structured support system would improve, strengthen and sustain health workforces in remote regions (Fisher and Fraser 2010). Improved support from managers is likely to improve the social exchange relationship and can better meet an employee's expectation and the reciprocal obligations for their psychological contract. Often, improved support translates into employee loyalty and stronger commitment behaviours. This is consistent with the findings from the study where there was a statistically significant correlation between supportive management practices and the health professionals intention to remain in their remote workplace (Onnis 2015). For example, health professionals who expressed loyalty towards their manager also felt that their manager provided for their personal safety, had adequate access to professional development opportunities, and had access to the resources they needed for their work.

Further, the findings suggested that supportive management practices provided a work environment conducive to positive individual wellbeing. Wellbeing has been linked to improved performance, a more positive outlook, and improves emotional adjustment of individuals to the work environment (Wright and Cropanzano 2004). Furthermore, health professionals who feel supported have better capacity to positively influence both their own personal wellbeing, as well as the health outcomes for remote populations. As these are all desirable outcomes for remote managers, it is reasonable to suggest that workforce sustainability can be improved through supportive, localised management practices. Therefore, management support is not only important for the health of your team and yourself; it promotes workforce sustainability, which benefits managers, health professionals, health service organisations and remote populations.

⟦⟧ Reflection
Take a moment to reflect on what support looks like in your remote workplace. Think about how you give and receive personal and professional support in your remote workplace.

1. How do you support the members of your team?
2. Is the support individualised or do you use the same methods to support each member of your team?
3. Does everyone need the same type of support?
4. Could you adapt your methods to suit each person, so that the support feels more personalised?
5. How do you contextualise your approach, do you need different ways to support your team in a remote workplace than you would use in an urban workplace? What is different?
6. Who supports you? Do you need to seek support or is it there whenever you need it?
7. If you are getting support from your team, think about this carefully. Should managers depend on their team for support? Where should managers be seeking support? Is there anything that you would change now that you have thought about who is supporting you?
8. Consider your earlier reflections about networks and peer support. How can they support you?

A HRM Framework for Remote Managers

An outcome of the study was an HRM framework developed specifically for remote managers. The intention was to identify which HRM concepts were influencing workforce sustainability, as well as the influence of management practices. Appendix A explains the research methodology and Appendix E explains how the data were analysed to develop the HRM framework. This chapter will explain and unpack the HRM framework and what it means for remote managers.

The *HRM Framework for a sustainable remote health workforce* (HRM-SRHW) (Fig. 8.1) is based on statistically significant findings that revealed the management practices and HR outcomes that influence the achievement of sustainable remote health workforces. The HRM-SRHW Framework is read from top to bottom. Starting at the top, the two management practices are leader-member exchange (LMX) and perceived organisational support (POS). LMX and POS were discussed in chapter six (see Social Exchange Theory in Appendix D). As a refresher, LMX relates to the quality of the employee-manager relationship and POS is the level of support the employee perceives is extended by their employer. These management practices influence the HR outcomes. HR outcomes are defined as the desired outcomes from the implementation of a particular HRM policy choice.

An example of this in practice: an organisation may decide that the HRM policy choice of HR Flow is consistent with their organisational values of 'leadership', 'promoting local employment' and 'growing their own workforce'. This policy choice may lead to a policy that creates pathways for internal promotion (e.g. succession planning, leadership programs). An HR outcome from this HRM Policy Choice (HR Flow) may be 'competence', 'cost-effectiveness' or 'commitment'. The implementation of this HRM policy choice may result in a health professional being more committed to an organisation that is facilitating their career path towards a management position, the targeted training may increase the competence of the health professional to perform their current and future roles, and the internal career paths may improve retention and reduce the costs of poor recruitment resulting in cost savings for the organisation. Therefore, the HRM Policy Choice results in desirable HR outcomes for the organisation, and improves workforce sustainability.

The HRM-SRHW highlights the most influential HR Outcomes for workforce sustainability. Further, the HRM-SRHW framework shows that where there are effective management practices, there are particular HR outcomes that assist in the

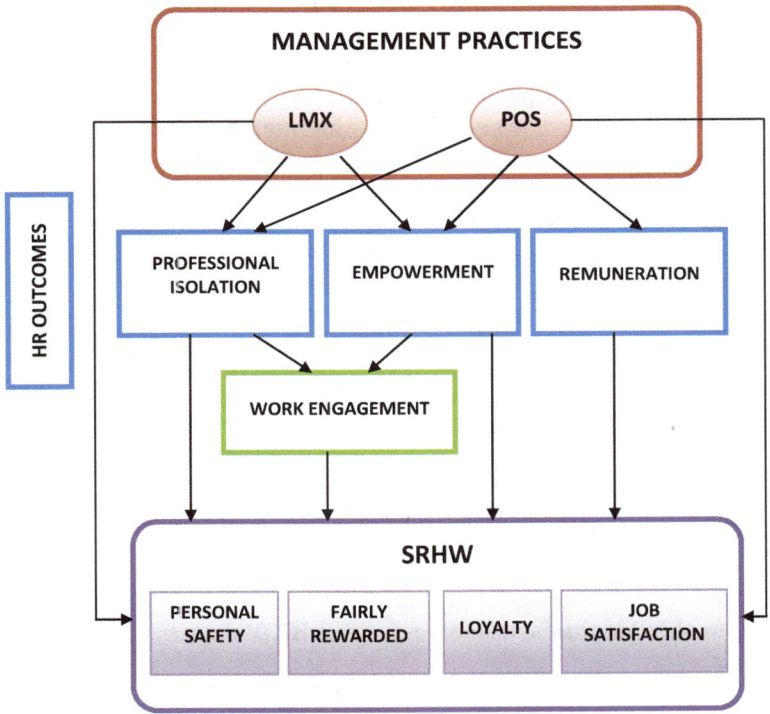

Fig. 8.1 HRM Framework for sustainable remote health workforces (HRM-SRHW)

achievement of the desired outcomes. Hence, effective management practices are essential. These management practices should support health professionals to prevent professional isolation and improve their sense of empowerment. When health professionals are empowered and have social exchange relationships that minimise professional isolation, the workforce will be more likely to be engaged. In the HRM-SRHW a sustainable remote health workforce (SRHW) is defined as a workforce where each individual employee feels that their employer provides for their personal safety, and rewards them fairly for the work that they do, for which they experience job satisfaction and remain loyal to their employer. Importantly, when the management practices were removed during statistical analyses, the relationships between the HR outcomes and a sustainable remote health workforce are no longer statistically significant, further highlighting the importance of effective management practices for workforce sustainability. The moderating effect of work engagement is consistent with the findings of O'Donohue et al. (2007, p. 307) who found that 'commitment to the profession underpinned high levels of job involvement not contingent on remaining with the organisation.' Global health workforce shortages, less job security, and increases in contingent labour and temporary employment suggest that occupational commitment will continue to increase as boundaryless and protean careers continue to rise.

The HRM-SRHW does not suggest cause-and-effect. This means that we can see the connections between specific HR outcomes and know that these HR outcomes are influencing remote workforce sustainability but we cannot say with certainty which ones are causing a particular observed phenomenon (e.g. voluntary turnover). This makes sense when considered within the complexity of employment relationships, where it is likely that a combination of factors contribute to decision-making. What we do know from the framework is that all of the HR outcomes in the framework had statistically significant relationships with the HR outcomes and management practices that are joined to each other by a line. Hence, management practices that focus on improving these HR outcomes, will lead to improved remote health workforce sustainability.

⊞ **Reflection**
Take some time to reflect on the HRM policy choices that you have implemented in your workplace.

1. How do they align with the HRM policy choices from the Harvard Analytical Framework for HRM (see Appendix B): HR Flow, Reward Systems, Employee Influence and Work Systems?
2. Do you find yourself implementing HRM policies from one particular HRM policy choice area more than another?
3. What implications does this have on the HR outcomes observed in your workplace?
4. How do the HRM policy choices that you implement align with your desired HR outcomes (commitment, competence, congruence, cost-effectiveness)? (see

Appendix B for a description of the HR Outcomes from the Harvard Analytical Framework for HRM)

5. Take an opportunity to reframe your mindset and start by thinking about the desired HR Outcomes in your workplace. Which HRM policy choices lead to these outcomes? What actions can you take to get your desired HR Outcomes?

HR Outcomes

In the HRM-SRHW Framework the significant HR outcomes for remote managers are: professional isolation, empowerment and remuneration. Professional isolation includes activities that reduce the health professional's sense of professional isolation, such as access to professional development and peer support that a health professional can experience due to their geographical isolation. Empowerment in the workplace was related to aspects of the work role, such as role clarity, decision-making and the extent to which employees have an increased scope for autonomous practice and decision making (Gould-Williams and Davies 2005). Remuneration included remote incentives, and extrinsic financial rewards. Remuneration is a contested area of benefit for remote workforces. There is evidence emphasising the benefits of incentives and bonus schemes for rural and remote recruitment and retention programs (Humphreys et al. 2012; Russell et al. 2013). In contrast, there is evidence suggesting that financial rewards are short-term motivators and as such do not promote long-term solutions for workforce sustainability (Campbell et al. 2012; Hackman and Oldham 1976). In terms of remuneration, the limited influence line managers have on remuneration is reflected in the framework with a significant relationship existing between remuneration and POS, but not between remuneration and LMX. This suggests that remote health professionals hold the organisation accountable for remuneration. This is indicative of the hierarchical structure of health systems, industry-based remuneration (e.g. Remote Area Nurses Incentive Package (RANIP)) and organisation-wide remuneration policies. This may also be a consequence of the absence of performance-based rewards. This separation of remuneration accountability to the employer level (organisation) appears to support the idea that health professionals can separate their dissatisfaction with work conditions and the satisfaction with the work role (Herzberg's Theory) which reinforces the notion that LMX, and the employee-manager relationship influence workforce sustainability. The study found that remote health professionals are motivated to remain in remote regions for reasons other than financial benefits and most health professionals suggested that these 'reasons' were either intrinsic rewards, or community/family connections to the remote region.

Effective management practices are those that localise HRM policy choices by interpreting and implementing them in a manner appropriate to the geographically remote location. Thus, the objective for a remote manager, is to effectively localise the implementation of HRM policies, to achieve desirable outcomes for the organisation, the employee, the community and themselves! Win-win-win-win. For further information on HR outcomes see the HRM theories and Harvard Analytical Framework for HRM information in Appendix B.

⛶ Reflection

Take some time to consider the influence of professional isolation, empowerment and remuneration on the sustainability of your workforce. Make a table with three columns similar to the example provided. In the first column write down the HR Outcomes. In the second column list the HRM Policies that you have implemented relevant to each HR Outcome. Then, in the third column list the management practices that would lead to your desirable outcomes. That is, what can you do tomorrow, next week, next month to implement HRM policies in a way that will most likely lead to the HR outcomes that you desire. When you have completed the third column highlight the management practices that you already do on a regular basis. The ones that are not highlighted are the management practices that you need to work on developing further.

HR outcomes	HRM policies	Management practices
Professional isolation	e.g. Professional development policy e.g. Conference attendance policy e.g. Mentoring and/or supervision policy	e.g. Demonstrate procedural fairness in allocating funding for professional development e.g. Ensure every team member attends one conference a year (equity)
Empowerment	e.g. Delegations policy e.g. Role description (level of responsibility)	e.g. Ensure that staff understand their role (role clarity) e.g. Be inclusive with decision-making, facilitating team input
Remuneration	e.g. Remote allowances e.g. Annual airfares	e.g. Balance team members' needs for leave and the health centre need for staff

Work Engagement

The recruitment advertising revealed that 'scope of practice' was frequently reported as one of the reasons that attracted current remote health professionals to remote regions. Therefore, when work engagement emerged in the HRM-SRHW it is hypothesised that work engagement is most likely the proxy for the 'commitment' HR outcome for remote health workforces. This provides insight into the turnover challenges being experienced in remote workplaces. With remote health professionals displaying commitment to their profession over commitment to the organisation; an opportunity exists for organisations to improve retention through management practices that support the health professional's career aspirations and professional development.

The Utrecht Work Engagement Scale (UWES) was used to measure work engagement. The UWES is the most widely used scale for measuring work engagement and has been used across diverse work groups and countries including translation into different languages (Mills et al. 2012; Nerstad et al. 2010; Seppala et al. 2009; Schaufeli et al. 2006). The work engagement questions from the UWES focus quite specifically on the work itself rather than organisational aspects of work. As health professionals working in remote regions are known to demonstrate high levels of work engagement (Opie et al. 2011). It is not surprising that work engagement emerged as one of the statistically significant HR outcomes. What is interesting is its associations with empowerment and professional isolation where work engagement moderated the relationship between management practices and a sustainable remote health workforce. This suggests that the HR Outcomes of professional isolation and empowerment will improve workforce sustainability, when the health professional shows certain levels of work engagement conducive to experiencing work satisfaction. In other words, we can surmise that when a remote manager effectively implements localised HRM policies, a health professional with a high level of work engagement, who feels empowered by their autonomous role and does not feel professionally isolated, is more likely to remain.

🛈 Resources

If you are interested in work engagement (an employee's level of engagement with the work that they do), rather than employee engagement (an employee's level of engagement with their employer, which usually includes organisational commitment) further information is available. The UWES is also available in languages other than English. The UWES Questions (English) are available at: https://www. wilmarschaufeli.nl/publications/Schaufeli/Tests/UWES_GB_9.pdf.

The UWES-9 questions were modified for the study and subsequently validated. The actual questions used in the study are available in the PhD Thesis, *A Sustainable Remote Health Workforce: Translating HRM Policy into Practice*. Available at: https://researchonline.jcu.edu.au/49756/.

An Integrated HRM Framework for Sustainable Remote Health Workforces

A figure introduced in chapter two showed the interrelatedness of the themes that emerged from a synthesis of the literature about the rewards and challenges of working in remote regions (Fig. 8.2). This figure provided a canvass on which to examine, compare and contrast the findings from the study, and to consider the relationship of these findings with those from other studies. This led to the development of the *Integrated HRM Framework for sustainable remote health workforces* (*I*-HRM-SRHW) (Fig. 8.3) which is the final HRM Framework that will be discussed in this book.

The four key HRM challenges and rewards discussed in section two were consistent with the HRM policy choices in the Harvard Analytical Framework for HRM (Beer et al. 1984). These were recruitment (HR flow); remuneration (rewards systems); relationships (employee influence); and resourcing (work systems) (see Appendix B for information about the Harvard Analytical Framework for HRM). Then, statistical data analysis revealed relationships among the HR outcomes and

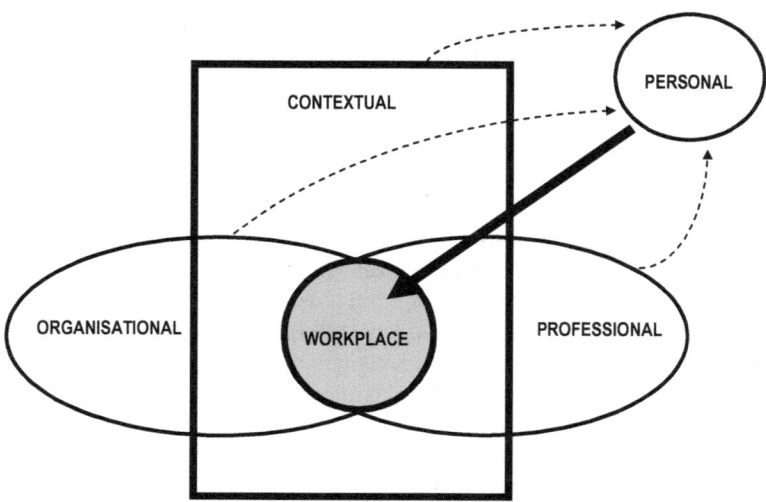

Fig. 8.2 Interrelatedness of themes describing key rewards and challenges for health professionals working in remote regions (the framework developed from this early conceptual version was published in Onnis and Pryce (2016, p. 47))

management practices that influence the attainment of sustainable remote health workforces (this lead to the development of the HRM-SRHW discussed previously). Next, the HRM-SRHW framework was combined with the findings from the synthesis of the literature to develop an *Integrated HRM Framework for sustainable remote health workforces* (*I*-HRM-SRHW) (Fig. 8.3). This integrated framework contributes to our understanding by combining what is already known through the health literature with HRM theories and remote workforce experiences. The *I*-HRM-SRHW is a valuable and new way for improving remote health workforce sustainability.

As explained in chapter two, the challenges and rewards of working in a remote workplace are shaped by contextual, professional, organisational and personal factors. These come together to create the 'localised context' for the remote workplace, depicted in grey in Fig. 8.2. In the *I*-HRM-SRHW (Fig. 8.3), the workplace contains the HRM-SRHW (Fig. 8.1), the framework that emerged from the study illustrating the influence of management practices and particular HR outcomes in the attainment of sustainable remote health workforces. Hence, the *I*-HRM-SRHW is a synthesis of the theoretical and empirical evidence from health and management research.

⑪ Reflection

Revisit the reflection activity from chapter two where you mapped each team member on Fig. 2.1 (Fig. 8.2 in this chapter). Think about where you placed each person in regard to person-fit in your remote workplace and consider whether this makes a difference in relation to the influence of HRM policies and HR Outcomes. Do you notice any patterns? Make a note of anything that you notice that may help you to adapt your management practices to improve workforce sustainability.

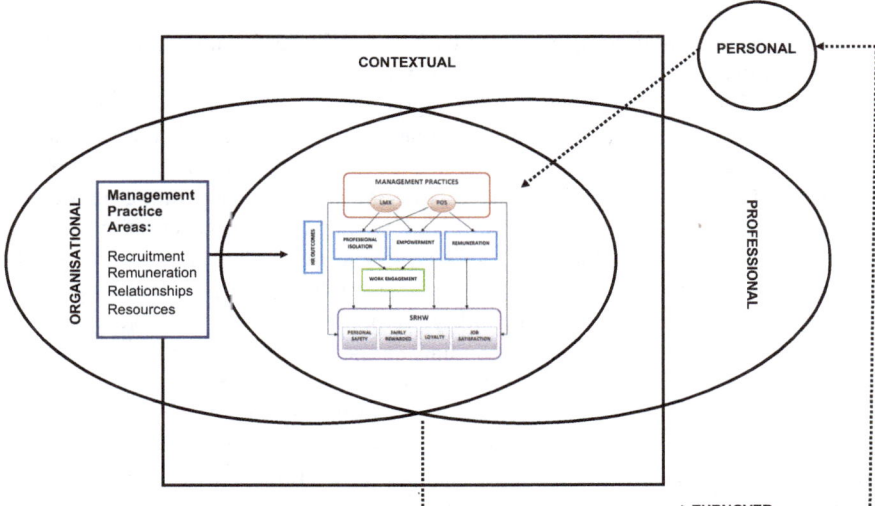

Fig. 8.3 Integrated HRM framework for sustainable remote health workforces (-HRM-SRHW)

For example, perhaps you noticed that the team member that was contained wholly within the workplace is motivated by HRM policies that increase remuneration. Perhaps you noticed that the team member that was contained mostly within the professional development section is motivated more by HRM Policy Choices that support their career development. When you look at these two examples, you would not use the same type of management practices to get the best performance out of each team member. The management practices would need to be adapted to both the context, and the individual team members. Over time, remote managers will become more astute at recognising when to use particular management practices and strategies when implementing HRM policies in their workplace if they wish to improve workforce sustainability.

How Can the *I*-HRM-SRHW Assist Remote Managers?

The *I*-HRM-SRHW shows that sustainable remote health workforces are influenced by factors associated with professional isolation, empowerment, work engagement and remuneration. Building on the theoretical foundations formed decades ago, the study provided further evidence of the critical role that managers play in workforce sustainability. Not all of the HRM concepts that appeared relevant for remote health workforces in an initial scoping review of the management literature were found to be statistically significant during data analysis. Therefore, there are many HR concepts that remain relevant to creating productive and healthy workplaces; however, the ones that comprise the *I*-HRM-SRHW were found to have a statistically significant relationship with management practices and workforce sustainability. This book has identified the key ingredients needed for sustainable remote health workforces; however, the remote manager needs to determine the combination and quantity of these ingredients needed for their particular remote context.

The practical application of the *I*-HRM-SRHW will differ depending on the type of health service or the geographically remote location; however, it provides a framework to guide the achievement of sustainable remote health workforces. The management practices, comprising of LMX and POS, were important aspects of remote health workforce sustainability, emphasising the influence of the employee-manager relationship (LMX) and the perception that the organisation is supportive of employees and their role in remote regions (POS). The quality of the employee-manager relationship will have a significant impact on workforce sustainability (Brunetto et al. 2010; Onnis 2014; Xerri 2013). Therefore, the manager, drawing on their skills and experience, becomes transformative as they master the 'science of managing people' in the remote context.

References

Beer M, Spector B, Lawrence PR, Mills DQ, Walton RE (1984) Managing human assets. Free Press, New York

Birks M, Mills J, Francis K, Coyle M, Davis J, Jones J (2010) Models of health service delivery in remote or isolated areas of Queensland: a multiple case study. Aust J Adv Nurs 28(1):25–34

Buckingham M, Coffman C (1999) First, break all the rules. What the worlds greatest managers do differently. Simon and Schuster, London

Brunetto Y, Farr-Wharton R, Shacklock K (2010) The impact of supervisor-subordinate relationships on morale: implications for public and private sector nurses' commitment. Hum Resour Manag J 20(2):206–225. https://doi.org/10.1111/j.1748-8583.2009.00117.x

Campbell N, McAllister L, Eley D (2012) The influence of motivation in recruitment and retention of rural and remote allied health professionals: a literature review. Rural Remote Health 12:1900

Fisher KR, Fraser JD (2010) Rural health career pathways: research themes in recruitment and retention. Aust Health Rev 34(3):292–296

Gould-Williams J, Davies F (2005) Using social exchange theory to predict the effects of HRM practice on employee outcomes. Pub Manage Rev 7(1):1–24

Hackman JR, Oldham GR (1976) Motivation through the design of work: test of a theory. Organ Behav Hum Perform 16(2):250–279. https://doi.org/10.1016/0030-5073(76)90016-7

Humphreys JS, McGrail MR, Joyce CM, Scott A, Kalb G (2012) Who should receive recruitment and retention incentives? Improved targeting of rural doctors using medical workforce data. Aust J Rural Health 20(1):3–10. https://doi.org/10.1111/j.1440-1584.2011.01252.x

Lenthall S, Wakerman J, Opie T, Dollard M, Dunn S, Knight S, MacLeod M, Watson C (2009) What stresses remote area nurses? Current knowledge and future action. Aust J Rural Health 17(4):208–213

Mills M, Culbertson S, Fullagar C (2012) Conceptualizing and measuring engagement: an analysis of the utrecht work engagement scale. J Happiness Stud 13(3):519–545

Nerstad C, Richardsen A, Martinussen M (2010) Factorial validity of the Utrecht Work Engagement Scale (UWES) across occupational groups in Norway. Scand J Psychol 51 (4):326–333

O'Donohue W, Donohue R, Grimmer M (2007) Research into the psychological contract: two Australian perspectives. Hum Res Dev Int 10(3):301–318

Onnis L (2014) Managers are the key to workforce stability: an HRM approach towards improving retention of health professionals in remote northern Australia. In: Proceedings, 28th Australia and New Zealand Academy of Management (ANZAM) conference. UTS Sydney

Onnis L (2015) An examination of supportive management practices promoting health workforce stability in remote northern Australia. Aust Psychiatry 23(6):679–682

Onnis L (2016) A sustainable remote health workforce: translating HRM policy into practice. Ph. D. Thesis. James Cook University

Onnis L, Pryce J (2016) Health professionals working in remote Australia: a review of the literature. Asia Pac J Hum Resour 54:32–56

Opie T, Lenthall S, Wakerman J, Dollard M, MacLeod M, Knight S, Rickard G, Dunn Sandra (2011) Occupational stress in the Australian nursing workforce: a comparison between hospital based nurses and nurses working in very remote communities. Aust J Adv Nurs 28(4):36–43

Purcell J, Hutchinson S (2007) Front-line managers as agents in the HRM-performance causal chain: theory, analysis and evidence. Hum Resour Manage J 17(1):3–20

Robertson R (2014) Leading on the edge. Wiley, Australia

Russell DJ, Wakerman J, Humphreys JS (2013) What is a reasonable length of employment for health workers in Australian rural and remote primary healthcare services? Aust Health Rev 37 (2):256–261

Schaufeli WB, Bakker AB, Salanova M (2006) The measurement of work engagement with a short questionnaire: a cross-national study. Educ Psychol Measur 66(4):701–716

Seppala P, Mauno S, Feldt T, Hakanen J, Kinnunen U, Tolvanen A, Schaufeli W (2009) The construct validity of the utrecht work engagement scale: multisample and longitudinal evidence. J Happiness Stud 10(4):459–481

Wright TA, Cropanzano R (2004) The role of psychological well-being in job performance: a fresh look at an age-old quest. Org Dyn 33(4):338–351

Xerri M (2013) Workplace relationships and the innovative behaviour of nursing employees: a social exchange perspective. Asia Pac J Hum Res 51(1):103–123

Practice Is Everything

9

> *In theory there is no difference between theory and practice. In practice there is.*
>
> Lawrence Peter 'Yogi' Berra, Professional Athlete, Manager.

Key Messages

- Events highlighting the continued complexity of managing health professionals in remote regions reinforce the importance of effective localised management practices.
- Despite the challenges there are serious implications for not improving and localising management practices; therefore, change is inevitable.
- For HRM policies to be effective, they must be localised, provide for patient care, and be flexible enough to promote safety in the complexity of geographically remote contexts.
- The remote manager's journey necessitates continued skill development and collective wisdom from real life experiences.

Why Practice Is Everything

'Empathy—like curiosity—requires attention. Attention requires time. When time is perceived to be scarce, empathy is diminished. This, in turn, diminishes the more human elements of business. Over time, an enterprise becomes one big machine, largely running on autopilot, like a soulless automaton. Good people leave bad managers, and talent flocks to more purpose-driven enterprises that are yet to succumb to the necromancy'. (Fox 2018)

This final chapter is a call to action. This book has offered ideas, suggestions for management tools and resources, opportunities to reflect on current practices in your remote workplace and a framework tailored to the needs of remote managers seeking workforce sustainability. Yet, nothing will change from reading this book unless

© Springer Nature Singapore Pte Ltd. 2019
L. Onnis, *HRM and Remote Health Workforce Sustainability*,
Management for Professionals, https://doi.org/10.1007/978-981-13-2059-0_9

your start to make changes in your management practices. If you have been doing the activities as you read the book, you may have started to see changes in your management practices and the ways in which HRM policies are implemented in your workplace. This is a great start but stay with it as new skills are needed as new situations emerge. As health professionals you work with people, as managers you work with people as well. Anyone who works with people knows that just when you think you have seen everything... something new comes along. In contemporary workplaces, the pace of change suggests that learning new skills, together with connected ways of working with others will be essential to thrive at work!

The previous chapter introduced the *I*-HRM-SRHW, a customised framework for remote health managers, and explained how HRM policy choices and localised management practices can positively influence remote health workforce sustainability. The focus on achieving workforce sustainability is not an end in itself, it is a contributing factor to improving the quality of health services. Hence, sustainable workforces contribute to improving the wellbeing of health professionals as well as access to health services for people living in remote regions. This serves as a reminder that positive health outcomes for remote populations are dependent on the implementation and localisation of HRM policies in all health service organisations.

Context Is Everything

The *I*-HRM-SRHW is customised for remote managers. The input from those on the ground was essential to make the *I*-HRM-SRHW framework relevant and applicable to remote managers. This is important because context is everything when it comes to the application of many things, including management practices. As remote managers you know the context within which you and your team work. Remote managers are well placed to localise implementation of HRM policies. However, the application of HRM policies is often hindered by organisational cultures, legislation, policies, senior management decisions, climate, resources, and environmental events that are beyond your control. Often, they may be beyond anyone's control. However, sometimes the application of HRM policies is often hindered by people and events that you can influence. Therefore, the message is influence what you can, as a manager know when to *do things right*, and as a leader know when to *do the right thing*.

The Contextual Challenges of Putting HRM Policies into Practice

In Australia, in 2016 there were several events that highlighted the continued complexity of managing health professionals in Australia's remote regions reinforcing the benefit of management practices that support effective localised implementation of HRM policies in remote regions. These events highlight the necessity for localised HRM polices that support remote health professionals commensurate with the environment in which they work.

Sadly, in March 2016, Gayle Woodford, a Remote Area Nurse (RAN) in the remote community of Fregon (South Australia) was tragically murdered when attending a night call out. The tragedy was described as 'a "wake-up call" for the industry' with Gayle Woodford's death highlighting 'the vulnerability of those who work alone in isolated areas' (Scopelianos 2016). Then in May 2016, the media reported that teachers in the remote community of Aurukun (Far North Queensland) had been evacuated over safety concerns. Towards the end of the article, in large letters it stated 'Health workers to remain in Aurukun', going on to explain that 'all eight nurses and four local health workers based in Aurukun had expressed a willingness to stay' even though the teachers were evacuated. The relevant Hospital and Health Service said that 'all our staff have duress alarms' and 'CCTV cameras monitor the clinic from Weipa' (a remote town approximately 200 km away with a driving time of around 2–4 hours depending on road conditions) (Aurukun Teachers Evacuated Again Over Safety Concerns 2016). On both of these occasions the respective Hospital and Health Services responded with descriptions about the way in which they will review policies or will afford support that will provide for the safety of their employees. The second article provides evidence of a common situation, where health professionals continue to work in potentially dangerous situations (e.g. conflict zones, pandemics) where they see that they, as health professionals, are needed. This emphasises the importance of localised policies congruent with the professional values and philosophies of health professionals working in remote regions, which may be at odds with city-centric HRM policies.

The geographical environment provides additional challenges for remote managers, illustrated in the tragic account of the events that transpired in the days and hours leading up to the passing of a remote nurse while on duty in Western Australia. Linton (2016, p. 4) reported that:

> '… nurses working at single nurse posts were rostered to work in the clinic between 8.00am to 5.00pm Monday to Friday. The clinic was closed for an hour for a lunch break, except for emergencies. Outside of these hours, the nurse was on call seven days per week for emergencies only within the town.'

This suggests that despite prescribed clinic hours the employer is aware that at a single nurse post, the nurse will be required to work after hours and be on call 24 hours a day, seven days a week. In the investigation, the employer reported that records showing the amount of hours that nurses actually work is incomplete because the HRM policy for an annual allowance for overtime, made it unlikely that nurses recorded the actual hours of overtime worked. The events, as described in the coroner's report, led to two remote nurses from two different remote communities doing a 'halfway meet[1]' which was against the employer's policy; however,

---

[1] A 'halfway meet' is where two health professionals drive towards each other, and meet at a point approximately halfway between the two health centres. This is usually done to share the burden of travel in an emergency situation, so that the patient can get to medical assistance as soon as possible.

in the given situation it was necessary. The 'halfway meet' was a known issue and the manager had drafted site instructions to address this issue, and the draft had been circulated for feedback but was not formalised. In this case, the manager had attempted to create policies; however, implementation appears to not have been effective. Regardless, the coroner's comment suggests that the policy would be irrelevant as nursing practices will always put the health of the patient first and therefore, the remote context necessitates practices that are often incompatible with HRM policies:

> 'I note that on this occasion, the halfway meet was only required because of a failure of the generator at the airstrip, which was not able to be quickly repaired. Even with the changes to the WACHS [Western Australia Country Health Service] policies about halfway meets and night driving, if a similar situation arose it would probably necessitate the nurse having to drive on the road at night. Events such as this will occur from time to time, given the remoteness of the region, and it is not a criticism of the WACHS to say that their policies will not be able to cover every eventuality and always guarantee a safe working environment for their remote area nurses.' (Linton 2016, p. 22)

In the most tragic of circumstances, this Coroner's report suggests that the policies designed to provide a safe workplace for remote nurses will not be enacted if the needs of their patients necessitate actions that contravene these HRM policies. Thus, it becomes clear that for HRM policies to be effective they must be localised, provide for patient care and be flexible enough to promote safety in the complexity of geographically remote contexts. Furthermore, it supports the findings from the study where occupational commitment appears to take priority over organisational commitment. As such, if clinical protocols take precedence over HRM policies, for HRM policies to translate into the intended work practices, effective management practices are important in policy development, interpretation and implementation. Considering the previous example, if 'halfway meets' are inevitable, a policy from head office prohibiting them is less likely to be as effective as a localised policy that provides for their safety should circumstances necessitate a 'halfway meet'.

[❗] Reflection

Consider the HRM policies in place to support and provide a safe work environment in your remote workplace.

1. Are they localised to your remote environment or are the generic health and safety policies?
2. How do you apply these policies in your remote workplace? Is there a need for adaptation so that they can actually provide a safe workplace for your team? (e.g. do they refer to equipment that you do not have in your remote workplace?)
3. Consider any HRM policies that you have localised to your remote workplace. How were they communicated to your remote team? How were they communicated to your organisation outside of the remote workplace? What worked? What can be improved?
4. Are there any HRM policies that are ineffective in your remote workplace in their current format?

Has this reflection identified any actions that are needed? Write down the actions that need to be taken and assign a completion date and a responsible person to each action.

🧰 Manager's Toolkit

Once you have worked out how best to localise the implementation of HRM policies that support the health and safety of your team; make notes or create a list of what works and put it in your Toolkit.

Also, revisit the list of contacts that you compiled in chapter seven about health, safety and wellbeing contacts for you and your team. Update this list and add any additional information that relates more specifically to localised management practices and local support services.

Managers Are the Key to Workforce Sustainability: The Next Generation

In an article published in 'The Australian' newspaper, Dr Ernest Hunter, a psychiatrist with more than 25 years experience working in remote communities, reflected on the changes to health services saying that 'turnover of non-functioning (and sometimes functioning) agencies undermined corporate memory and institutional learning' referring to a paper published to document the history of a remote mental health team that had experienced relative stability when compared to similar mental health services operating in comparable contexts (Hunter et al. 2013; Hunter 2015). Hunter (2015) believed that it demonstrated what can be achieved with management support and commitment. Hunter's (2015) proposal that relationships are essential for the provision of health services is echoed in this book, where relationships are essential for workforce sustainability.

This book has argued that managers are the key to remote health workforce sustainability. Some remote health professionals predict that the distance between management and remote health professionals will widen with the direction in which both government policies and organisational HRM policies are heading (Hunter 2015). However, others report that this serves to highlight the challenges and implications for not improving and localising management practices; hence, change is inevitable (Wakerman and Davey 2008). As Wakerman and Davey (2008, p. 17) conclude in their case study about the experiences of a RAN, 'There is a new era of nurses coming... the next generation is not going to put up with this...' Hence, a solutions-based approach is essential.

In particular, current remote managers should consider how they are contributing to improving the transition from clinician into management for aspiring remote managers. The *I*-HRM-SRHW suggests that localised management practices associated with professional isolation, and empowerment will be effective. The role of the remote manager in preparing the next generation of remote managers is vital for attaining workforce sustainability and delivering quality health services. In fact, it is essential in contemporary remote workplaces where people leave managers.

The Remote Manager's Journey Continues

The book commenced with brief description of management and leadership and some comforting words to remind remote managers that they are not expected to know what to do in every situation. So as the book comes to a close, it is important to revisit these ideas. Managers do not have the answer to every question and cannot be expected to universally know what to do. However, managers do need to be able to 'manage' any situation that arises. This means that remote managers need to know when to seek assistance, when to rely on the skills and expertise of others, when to take the lead, and when to be quiet and do nothing so that others can learn. As a manager you must be a leader in your workplace. As many accomplished business leaders and academics have proclaimed, not all leaders are managers, but all managers must be leaders to be successful.

By now, you understand that for the remote manager there is no 'checklist' or no one best way when it comes to managing people. By completing the activities and reflections in this book you have started to develop skills in localising the implementation of HRM policies. In doing this you may have started to realise that the remote manager's journey, like most professions, necessitates a continual journey of skill development and an accumulation of wisdom arising from real life experiences. There is a wide assortment of books, podcasts, blogs, articles and training courses that can help you to develop and fill your Manager's Toolkit with tools and resources to help you to thrive in remote regions just like the Boab of the Kimberley (Chap. 1).

The following section contains some suggestions for further reading and online resources to help you on your way.

References

Aurukun Teachers Evacuated Again Over Safety Concerns (2016, May 26) ABC News. Retrieved from https://www.abc.net.au/news/2016-05-25/aurukun-teachers-evacuated-for-second-time/7444630. Accessed 16 Oct 2018

Fox J (2018, March 28) The 5 Pernicious Patterns Most Leaders Cannot See. The hidden forces that sap our ability to innovate & change. Available at https://www.drjasonfox.com/blog/5-hidden-patterns. Accessed 16 October 2018

Hunter E, Onnis L, Santhanam-Martin R, Skalicky J, Gynther B, Dyer G (2013) Beasts of burden or organised cooperation: the story of a mental health team in remote Indigenous Australia. Australas Psychiatry Australa Psychiatry 21(6):572–577

Hunter E (2015, November 28). Health sectors puruit of efficiency comes at great cost. The Australian. Retrieved from http://www.theaustralian.com.au/news

Linton SH (2016) Inquest into the death of Gonda (aka Connie) SMITH (11025/2012). http://www.coronerscourt.wa.gov.au/_files/Smith%20(Gonda)%20finding.pdf. Accessed 2 June 2018

Scopelianos S (2016) Gayle Woodford: man charged with murdering 'popular' outback SA nurse. ABC News. http://www.abc.net.au/news. Accessed 5 Sept 2017

Wakerman J, Davey C (2008) Rural and remote health management: 'the next generation is not going to put up with this…'. Asia Pacific. J Health Manage 3(1):13–18

Appendix A

A.1 Research Study Methodology

The research was conducted in tropical northern Australia which included the Kimberley (Western Australia), the Top End (Northern Territory) and Far North/North West Queensland. A mixed methods research methodology offered a robust method to integrate both qualitative and quantitative data. There were three data sources: (1) semi-structured interviews (qualitative); (2) questionnaires (qualitative and quantitative); and (3) recruitment advertising (qualitative and quantitative).

A.1.1 Interviews

A purposive sampling method of snowballing was used to recruit participants for the interviews. The semi-structured approach used a set of questions to guide the interviews improving the consistency in the way the questions were asked (see Onnis 2016 for the interview questions). A thematic analysis of the transcripts (n = 24) was conducted using NVivo 10 (QSR International Pty Ltd, Melbourne, Australia). Further information about the data analysis techniques is available in Onnis 2016.

A.1.2 Questionnaire

A probability sampling method was used to obtain the remote health professional's perspective through an online questionnaire. The questionnaire was first distributed from January–July 2014, and then the same questionnaire was distributed from January–July 2015 to eight organisations (two government health departments, two Aboriginal Community Controlled Health Organisations (ACCHOs), two non-profit organisations, and two recruitment agencies). The questionnaire contained 60 Likert scale questions, five short answer text questions, and questions collecting demographic data. The link to the online questionnaire was

© Springer Nature Singapore Pte Ltd. 2019
L. Onnis, *HRM and Remote Health Workforce Sustainability*,
Management for Professionals, https://doi.org/10.1007/978-981-13-2059-0

sent to health professionals working in remote regions through their internal email system by HR or a senior manager. This method of distribution enabled a high level of confidentiality to be offered, with the researcher unaware of the potential participant's names, and the organisation unable to access the completed questionnaires. There were 213 completed questionnaires analysed. There is further information about the data analysis methods in Appendix E. Also, additional information about the data analysis techniques and a copy of the questionnaire is available in Onnis (2016).

A.1.3 Recruitment Advertising

Recruitment advertisements (n = 3311) from five recruitment websites were analysed from August 2013 to July 2015. These websites were: Western Australia Government website (www.jobs.wa.gov.au); Northern Territory Government website (www.jobs.nt.gov.au); Queensland Government website (www.smartjobs.qld.gov.au); Seek (www.seek.com.au); and CareerOne (www.careerone.com.au). Content analysis was conducted to systematically analyse the text. Content analysis enabled the written data to be coded and then counted and analysed using descriptive quantitative data analysis techniques. Descriptive data analyses including frequencies and cross-tabulations were conducted using the statistical software package SPSS22. Further information about the data analysis techniques is available in Onnis (2016). The advertisement was selected if it met the following criteria:

1. The position was in a region included in the study
2. The position involved contact with patients for treatment or to enable/assist patients to receive treatment; or the management of people who had contact with patients for treatment or to enable/assist patients to receive treatment
3. They required a health-related qualification and/or experience in a role that provided healthcare services (as described in criterion two).

Reference

Onnis L (2016) A Sustainable Remote Health Workforce: Translating HRM Policy into Practice. Ph.D. Thesis. James Cook University

Appendix B

B.1 Introduction to HRM

HRM is about the systems that support managers in managing people at work. HRM practices are distinguished by whether the focus is on the *humans* or the *resources*—referred to as *hard* or *soft* HRM (Guest 1987; Truss 1997). The 'hard' side of HRM considers employees to be costs associated with doing business whereas 'soft' HRM considers employees as resources that provide the organisation with a competitive advantage (Gill and Meyer 2011). The differentiation between 'soft' and 'hard' HRM becomes apparent in the implementation of HRM policies and management practices. For example, soft HRM policies support the development of organisational commitment, loyalty and can be seen in policies which provide flexibility and build the foundation for the desired organisational culture. In contrast, hard HRM policies and practices contribute to the governance and compliance systems and can be seen in policies such as performance-based rewards and remuneration packages offered to employees. Often organisations promote the soft commitment model to build desired culture, yet, in reality employees experience control similar to the hard model. Inconsistent interpretation of HRM policies may lead to management practices that not only confuse the HRM policy message but result in miscommunication that is detrimental for organisations, such as turnover and low job satisfaction (Gill and Myer 2011; Townsend et al. 2012).

To improve HRM policy outcomes it is important to recognise that HRM policies and practices make up only one set of the many signals that remote managers receive daily (Townsend et al. 2012). Managers must decipher these signals to determine the best actions for their context. That is, on any given day, managers have HRM policy signals, as well as clinical policy signals, financial policy signals, governance policy signals, and many other signals. Therefore, effective communication is vital as HRM policies are not the only policies that frontline managers interpret and often, in the health sector, policies for patient care and budgetary issues are given higher priorities by managers. As employee perceptions of HRM practices are usually the practices applied by managers it is clear that managers play a central role in communicating policies, and may be

© Springer Nature Singapore Pte Ltd. 2019
L. Onnis, *HRM and Remote Health Workforce Sustainability*,
Management for Professionals, https://doi.org/10.1007/978-981-13-2059-0

selective in determining which policies are communicated using their discretionary power to determine which information is important and how it will be disseminated (Townsend et al. 2012). Until managers see the benefit of the HRM policies, it is unlikely that they will be prioritised.

HRM describes the people management practices that support managers to implement systems and work practices that meet the needs of their clients, the health service, and the health professionals that they manage within an uncertain and continuously evolving environment. As such, theories that can inform managers·about why some practices may be more successful than others in given situations are valuable for the remote manager.

B.1.1   Harvard Analytical Framework for HRM

The Harvard Analytical Framework for HRM comprises four HRM policy choices that encompass the decisions about implementing specific HRM policies and procedures in an organisation. Each of the four HRM policy areas is characterised by actions that lead to the HR outcomes.

B.1.1.1   HRM Policy Choices

HR Flow

In a dynamic environment, organisations must focus on 'managing the flow of people in, through, and out of the organization' (Beer et al. 1984, p. 66). There are many internal factors (e.g. management practices) and external factors (e.g. industry skills shortage) that impact HR Flow. However, HR Flow policy choices are in fact closely related to all other HRM policy choices which endeavour to achieve workforce sustainability. Before investing resources in retention initiatives, organisations need to understand the nature of turnover in their particular context, otherwise is unlikely to maximize their return on the investments (Maertz and Boyar 2012). For organisations with limited resources, it makes sense to focus attention on assessing the attachment and withdrawal behaviour of current employees, those employees whose decisions can still be influenced by management (Maertz and Boyar 2012).

Reward Systems

The primary focus for most organisations is to design a job so that employees work efficiently receiving an appropriate level of compensation (Giancola 2011). Intrinsic motivation is described as the 'doing of an activity for its inherent satisfactions rather than for some separable consequence' (Ryan and Deci 2000, p. 56), in other words, the reward is in the activity itself (Baard et al. 2004; Kanungo and Hartwick 1987). Intrinsic rewards include: promotion, authority, responsibility, and participation in decision making, praise from supervisors, praise from co-workers, recognition, and awards for superior performance (Giancola 2011; Kanungo and Hartwick 1987). In contrast, extrinsic motivation is when 'an activity is done in order to attain some separable outcome' (Ryan and Deci 2000, p. 60). Extrinsic rewards are often monetary, such as pay and bonuses (Giancola 2011). Economic principles prescribe to the notion that people respond to incentives; however, these incentives which are usually in the form of rewards and punishments are often counterproductive and undermine intrinsic motivation (Gagné and Deci 2005; Ryan and Deci 2000).

Employee Influence

Employee influence describes how employees can 'act to improve or protect their economic share, psychological satisfaction, and rights' at work including how this influence is exercised (Beer et al. 1984, p. 40). Employee influence includes opportunities for employee participation in decision-making, practices which improve employee morale and job satisfaction (Bhattacharya and Wright 2005). Organisations that enable employee participation in decision-making, are regarded as more transparent and treat employees with respect strengthening perceptions of congruence between employee and organisational values which integrates them into the organisation, and enhances commitment (Wright and Kehoe 2008). Furthermore, interpersonal relationships and social interactions accelerate feelings of inclusion, social cohesion and organisational embeddedness thus improving organisational commitment (Wright and Kehoe 2008).

Work Systems

Management practices in implementing work systems will have a strong effect on the organisation's effectiveness (Beer et al. 1984, p. 153). While employees work within the system, the creation and continuation of the work systems are the responsibilities of management. Therefore, for employees most performance variation is probably due to system factors beyond their control; however, for the managers who are responsible for the development and maintenance of the systems it is not always clear where their control lies and how or when it should be exercised (Waldman 1994). Where employees feel a sense of freedom and power to influence the system they feel autonomous in their work environment. Furthermore, it enables them an opportunity to modify themselves or the work environment enhancing person-fit and may 'serve to moderate the extent to which individuals are able to significantly influence a system' (Waldman 1994, p. 527).

People interact with systems so even 'if a system is invariant, people may react to it differentially because of their different abilities, values [and] expectations' (Waldman 1994, p. 517). For example, inconsistent leadership or management practices can be a source of variation within a system (Waldman 1994). People can experience dissatisfaction with a system despite a favourable outcome. This may be due to a perception of injustice in the process or system. On the other hand, 'the use of a process viewed as fair and just can make negative outcomes more palatable'

(Ko and Hur 2014, p. 179). Thus, even when there are low levels of satisfaction with employment benefits, if there is high managerial trustworthiness employees remain intrinsically motivated (Ko and Hur 2014, p. 180).

B.1.1.2   HR Outcomes

HRM policy choices are not isolated policy or practice decisions; they are part of the broader organisational policies and impact the organisation and society on varying levels. The HRM policy choices and implementation methods can lead to one or more the four HR outcomes.

Commitment

HRM policies influence employee commitment behaviours. 'Increased commitment can result not only in more loyalty and better performance for the organization, but also in self-worth, dignity, psychological involvement, and identity for the individual' (Beer et al. 1984, p.19). Commitment is associated with an exchange relationship. Employees can be committed to their profession, their colleagues (team) or their organisation (Irving et al. 1997). Both organisational and occupational commitment influence HR outcomes, such as turnover and intention to leave (Gambino 2010; Knights and Kennedy 2005).

Competence

Competence describes the extent to which HRM policies attract, and develop people with the skills and knowledge needed by an organisation. If the necessary skills and knowledge are available at the right time, both organisations and employees benefit through economic gains and an increased sense of self-worth (Beer et al. 1984). Hence, the relationship between competence and empowerment in the workplace is based on an assumption that empowerment is shaped by the work context (Spreitzer 1996).

Congruence

A lack of 'congruence can be costly to management in terms of time, money, and energy' (Beer et al. 1984, p. 19). Individual employees seek alignment between themselves and their work, as well as with the organisational and broader societal contexts. Therefore, 'individual employees were more satisfied and committed to the organisation when their values were congruent with that of their supervisors' (Rosete 2006, p. 8). This suggests that congruence between the organisation's values and those of the individual employee can impact job satisfaction, organisational commitment and intention to leave an organisation (Rosete 2006). Hence, recruitment and selection processes may be more successful if they focus more on attempting to fit individuals with managers, work groups and organisations, rather than fitting people to defined jobs.

Cost-effectiveness

The impact of the 'cost-effectiveness of a given policy in terms of wages, benefits, turnover, absenteeism, strikes, and so on' should be 'considered for organizations, individuals, and society as a whole (Beer et al. 1984, p. 19). An association between high commitment HRM models and the organisation's financial performance is established. More recently, social exchanges have been viewed from a cost–benefit perspective, similar to an economic exchange, with the exchange being intangible social costs and benefits instead of more tangible monetary gains (Cropanzano and Mitchell 2005; Xerri 2013).

References

Baard PP, Deci EL, Ryan RM (2004) Intrinsic need satisfaction: a motivational basis of performance and well-being in two work settings. J Appl Soc Psychol 34(10):2045–2068

Beer M, Spector B, Lawrence PR, Mills DQ, Walton RE (1984) Managing human assets. The Free Press, New York

Bhattacharya M, Wright P (2005) Managing human assets in an uncertain world: applying real options theory to HRM. Int J Hum Res Manag 16(6):929–948

Cropanzano R, Mitchell M (2005) Social exchange theory: an interdisciplinary review. J Manag 31(6):874–900

Gagné M, Deci EL (2005) Self-determination theory and work motivation. J Organ Behav 26(4):331–362

Gambino KM (2010) Motivation for entry, occupational commitment and intent to remain: a survey regarding registered Nurse retention. J Adv Nurs 66(11):2532–2541

Giancola FL (2011) Examining the job itself as a source of employee motivation. Compensation Benefits Rev 43(1):23–29

Gill C, Meyer D (2011) The role and impact of HRM policy. Int J Organ Anal 19(1):5–28

Guest DE (1987) Human resource management and industrial relations. J Manag Stud 24(5):503–521

Irving PG, Coleman DF, Cooper CL (1997) Further assessments of a three-component model of occupational commitment: generalizability and differences across occupations. J Appl Psychol 82(3):444–452

Kanungo RN, Hartwick J (1987) An alternative to the intrinsic-extrinsic dichotomy of work. J Manag 13(4):751–766

Knights JA, Kennedy BJ (2005) Psychological contract violation: impacts on job satisfaction and organizational commitment among Australian senior public servants. Appl H.R.M. Res 10(2): 57–72

Knights JA, Kennedy BJ (2005) Psychological contract violation: impacts on job satisfaction and organizational commitment among Australian senior public servants. Appl H.R.M. Res 10(2): 57–72

Ko J, Hur S (2014) The impacts of employee benefits, procedural justice, and managerial trustworthiness on work attitudes: integrated understanding based on social exchange theory. Publ Adm Rev 74(2) 176–187

Maertz CP, Boyar SL (2012) Theory-driven development of a comprehensive turnover-attachment motive survey. Hum Res Manag 51(1):71–98

Rosete D (2006) The impact of organisational values and performance management congruency on satisfaction and commitment. Asia Pac J Hum Res 44(1):7–24

Ryan RM, Deci EL (2000) Intrinsic and extrinsic motivations: classic definitions and new directions. Contemporary Educ Psychol 25:54–67

Ryan RM, Deci EL (2000) Intrinsic and extrinsic motivations: classic definitions and new directions. Contemporary Educ Psychol 25:54–67

Spreitzer G (1996) Social structural characteristics of psychological empowerment. Acad Manag J 39(2):483–504

Townsend K, Wilkinson A, Allan C (2012) Mixed signals in HRM: the HRM role of hospital line managers. Hum Res Manag J 22(3):267–282

Truss, C. (1997). Soft and Hard Models of Human Resource Management: A Reappraisal. J Manag Stud, 34(1), 53–73.https://doi.org/10.1111/1467-6486.00042

Waldman DA (1994) The contributions of total quality management to a theory of work performance. Acad Manag Rev 19(3):510–536

Wright PM, Kehoe RR (2008) Human resource practices and organizational commitment: A deeper examination. Asia Pac J Hum Res 46(1):6–20

Xerri M (2013) Workplace relationships and the innovative behaviour of nursing employees: a social exchange perspective. Asia Pac J Hum Res 51(1):103–123

Appendix C

C.1 Psychological Contract Theory

Psychological Contract Theory (PCT) describes an individual employee's beliefs about 'what they think they are entitled to receive because of real or perceived promises' from their employer (Bartlett 2001, p. 337). The employee's belief about the reciprocal exchange agreement that forms the employment relationship is unspecified and implicit, and thus not known to anyone other than the employee (Cullinane and Dundon 2006). These obligations may be transactional (e.g. pay) or relational (e.g. loyalty in exchange for job security). Transactional psychological contracts are characterised by specific, short-term, monetary obligations (Coyle-Shapiro and Kessler 2000; Thomas et al. 2003). In contrast, relational contracts emphasise broad, long-term, socio-emotional obligations, such as commitment, loyalty, fairness, trust, job security, role clarity, and career development (Coyle-Shapiro and Kessler 2000; Thomas et al. 2003). Where employees perceive that organisations value and treat them equitably, they will reciprocate with positive attitudes and behaviours (Gould-Williams and Davies 2005). However, if an employee perceives inequity in the distribution of rewards or injustice within the workplace, unmet expectations may be viewed as breaches or violations of the psychological contract (Knights and Kennedy 2005).

A psychological contract breach and violation are responses to unfulfilled obligations. A breach is described as the cognitive evaluation, that is, a mental calculation of what has been received compared to what the employee believes was promised, whereas, violation is the emotional response that may follow from the breach (Knights and Kennedy 2005; Zhao et al. 2007). In other words, a violation is an outcome of breach and the emotion of the violation is often translated into the behaviour that results in voluntary turnover (O'Donohue and Nelson 2007; Zhao et al. 2007). There are a variety of actions, expectations and complications to the psychological contract beyond the obvious difficulties associated with an unwritten contract developed by only one party. There may also be aspects of employment, such as temporary workers, occupations and hierarchies that influence psychological contract fulfilment. For example, Guest (2004) found that temporary employees were more likely than permanent employees to perceive their contracts

© Springer Nature Singapore Pte Ltd. 2019
L. Onnis, *HRM and Remote Health Workforce Sustainability*,
Management for Professionals, https://doi.org/10.1007/978-981-13-2059-0

as transactional rather than relational. Additionally, 'it is possible that employees from different occupations have different psychological contracts and react to breach in different ways' (Zhao et al. 2007, p. 671).

It is difficult to know how managers can foster a relationship conducive to maintaining psychological contracts, given that they are unwritten and formed by the employee; however, it is known that turnover intention is linked to psychological detachment from an organisation. Therefore, perceived psychological contract breach and violation reduces or eliminates such obligations which influences voluntary turnover so strengthening an employee's relationship with their manager should strengthen the connection of the employee to the organisation (Ko and Hur 2014; Knights and Kennedy 2005; Maertz and Boyar 2012).

C.1.1 Psychological Contact Theory and Social Exchange Theory

Psychological Contact Theory and Social Exchange Theory (SET) have similar underlying principals; for example, when an individual perceives the organisation as failing to fulfil its obligations the individual will change their behaviour and attitudes towards the organisation. Social exchanges and reciprocity play a critical role in the psychological contract where mutual obligations, such as social exchanges shape the formation of the psychological contract. Furthermore, PCT and SET are consistent with the high commitment approach to HRM which promotes the benefits of positive psychological links between employees and the employer (Gould-Williams and Davies 2005).

References

Bartlett KR (2001) The relationship between training and organizational commitment: a study in the health care field. Hum Res Dev Q 12(4):335–352

Coyle-Shapiro J, Kessler I (2000) Consequences of the psychological contract for the employment relationship: a large scale survey. J Manag Stud 37(7):903–930

Cullinane N, Dundon T (2006) The psychological contract: a critical review. Int J Manag Rev 8(2):113–129

Gould-Williams J, Davies F (2005) Using social exchange theory to predict the effects of HRM practice on employee outcomes. Public Management Review 7(1):1–24

Guest D (2004) Flexible employment contracts, the psychological contract and employee outcomes: an analysis and review of the evidence. Int J Manag Rev 5(1):1–19

Knights JA, Kennedy BJ (2005) Psychological contract violation: impacts on job satisfaction and organizational commitment among Australian senior public servants. Appl H.R.M. Res 10(2):57–72

Ko J, Hur S (2014) The impacts of employee benefits, procedural justice, and managerial trustworthiness on work attitudes: integrated understanding based on social exchange theory. Public Adm Rev 74(2):176–187

Maertz CP, Boyar SL (2012) Theory-driven development of a comprehensive turnover-attachment motive survey. Hum Res Manag 51(1):71–98

O'Donohue W, Nelson L (2007) Let's be professional about this: ideology and the psychological ontracts of registered nurses. J Nurs Manag 15(5):547–555

Thomas D, Au K, Ravlin E (2003) Cultural variation and the psychological contract. J Organ Behav 24(5):451–471

Zhao H, Wayne S, Glibkowski B, Bravo J (2007) The impact of psychological contract breach on work-related outcomes: a meta-analysis. Pers Psychol 60(3):647–680

Appendix D

D.1 Social Exchange Theory

Social Exchange Theory (SET) focuses on relationships and the perceived obligations which form exchange relationships in the workplace. When the employment relationship is viewed as an exchange it is often described as consisting of social and economic exchanges which include contractual arrangements enforceable through law, and social arrangements premised on the exchange of favours arising from a sense of obligation to the other party (Gould-Williams and Davies 2005). Social exchanges are more difficult to quantify, but are quite influential in the workplace, particularly when considering workplace relationships. These social exchanges involve interactions that generate obligations over a period of time with employees who are more satisfied with the outcomes of these exchanges in the workplace, more likely to respond with performance improvements (Xerri 2013).

There are two types of social exchange, one is perceived organisational support (POS) and the other is leader-member exchange (LMX) (Brunetto et al. 2016; Xerri 2013). Perceived organisational support focuses on an exchange relationship between an employee and an organisation whereas leader-member exchange emphasises the quality of exchange relationship between the employee and their manager. It is expected that high POS would result in lower turnover as reciprocity usually means that people feel obligated to help people who have helped them (Allen et al. 2003).

It is argued that the quality of the exchange between employees and their manager reflects a degree of trust, loyalty and respect (Brunetto et al. 2016; Ko and Hur 2014). Where there are good quality exchange relationships everyone enjoys the benefits of a cohesive team (Brunetto et al. 2016). LMX affects the level of job satisfaction as well as organisational commitment of employees (Brunetto et al. 2011). When employees perceive their manager as trustworthy, these beliefs can extend to the organisation; with employees more likely to work for an organisation they trust (Ko and Hur 2014). When an employee feels valued and supported, they perceive the benefit is in exchange for an attitude or behaviour on their part, thus reinforcing the reciprocal exchange relationship (Allen et al. 2003). Therefore,

© Springer Nature Singapore Pte Ltd. 2019
L. Onnis, *HRM and Remote Health Workforce Sustainability*,
Management for Professionals, https://doi.org/10.1007/978-981-13-2059-0

traditional benefits, which are those benefits available to all employees regardless of performance (e.g. Awards pay rates), are unlikely to be associated with POS (Ko and Hur 2014). Consequently, if organisations want 'to build positive social exchange with their employees, employee benefits should be recognized as benefits beyond those typically offered by most organizations' (Ko and Hur 2014, p. 183), such as performance-based rewards that are earned rather than automatically paid after a given period of time.

The presence of HRM policies and practices implies that an organisation cares about and values its employees. Specifically, there are benefits in an approach to people management that aims to enhance performance by empowering, developing and trusting employees to work confidently and competently on the basis of mutuality of interests (Gould-Williams and Davies 2005).

Social Exchange Theory offers a key theoretical evidence-base to examine the sustainability of a workforce because the 'exchange' and 'obligation' implies that the employee participates in the relationship. According to Allen et al. (2003) continued participation is a way in which employees repay the organisation, so POS can encourage continued membership in the organisation. Furthermore, it is the employee's understanding of the exchange relationship that defines 'the employment relationship and subsequently the psychological contract' (Allen et al. 2003, p. 103).

References

Allen DG, Shore LM, Griffeth, RW (2003) The role of perceived organizational support and supportive human resource practices in the turnover process. J Management 29(1):99–118

Brunetto Y, Farr-Wharton R, Shacklock K (2011) Using the harvard HRM model to conceptualise the impact of changes to supervision upon HRM outcomes for different types of Australian public sector employees. Int J Hum Res Manag 22(3):553–573

Brunetto Y, Xerri M, Trinchero E, Farr-Wharton R, Shacklock K, Borgonovi E (2016) Public–private sector comparisons of nurses' work harassment using set: Italy and Australia. Public Manag Rev 18(10):1479–1503

Gould-Williams J, Davies F (2005) Using social exchange theory to predict the effects of HRM practice on employee outcomes. Public Manag Rev 7(1):1–24

Ko J, Hur S (2014) The impacts of employee benefits, procedural justice, and managerial trustworthiness on work attitudes: integrated understanding based on social exchange theory. Public Adm Rev 74(2):176–187

Xerri M (2013) Workplace relationships and the innovative behaviour of nursing employees: a social exchange perspective. Asia Pac J Hum Res 51(1):103–123

Zhao H, Wayne S, Glibkowski B, Bravo J (2007) The impact of psychological contract breach on work-related outcomes: a meta-analysis. Pers Psychol 60(3):647–680

Appendix E

E.1 Developing the HRM-SRHW Framework

This section describes how the HRM framework for remote managers was developed. An analysis of the data was conducted using IBM SPSS22 and SPSS23. Then, IBM SPSS AMOS was used to conduct a path analysis. More detailed information about the data analysis methods, including the questions and statistical analysis is available in the Ph.D. thesis, *A Sustainable Remote Health Workforce: Translating HRM policies into practice* (Onnis 2016). However, a brief summary is contained in this section to provide background to the development of the HRM framework.

E.1.1 Identifying the Factors

The questionnaire data was assessed for normality through SPSS22 and factor analysis was conducted as a data reduction technique to determine whether the sixty questions could be summarised as a smaller set of factors (Tabachnick and Fidell 2007). Confirmatory Factor Analysis (CFA) using Principal Components Analysis identified ten factors which explained 63.76% of the variance. Varimax orthogonal rotation method with Kaiser Normalisation, resulted in a Kaiser-Myer Olkin Measure of Sampling Adequacy of 0.828 and a significant Bartlett's Test of Sphericity ($p < 0.001$). The Cronbach's alpha was calculated for each of the ten factors to determine internal reliability (Table E.1). Two factors (work conditions and embeddedness) with a Cronbach's alpha below 0.50 were removed. The remaining eight factors had a Cronbach's alpha ranging from 0.64 to 0.93, and while the preference for internal reliability was a Cronbach's alpha above 0.70, the majority of factors had a Cronbach's alpha above 0.70 so they were deemed to have adequate levels of internal consistency (Briggs et al. 2009; Pelletier et al. 1995).

© Springer Nature Singapore Pte Ltd. 2019
L. Onnis, *HRM and Remote Health Workforce Sustainability*,
Management for Professionals, https://doi.org/10.1007/978-981-13-2059-0

Table E.1 Factors identified through confirmatory factor analysis

Factor	Cronbach's Alpha
Leader member exchange (LMX)	0.930
Work engagement	0.879
Perceived organisational support (POS)	0.820
Empowered	0.771
Personal isolation	0.744
Professional isolation	0.663
Competence	0.636
Remuneration	0.631
Work conditions	0.500
Embeddedness	0.498

E.1.2 Describing the Key Factors

Data analysis proceeded with eight factors which comprised two management factors (LMX, POS), work engagement, and five HR outcomes (competence, professional isolation, empowerment, personal isolation and remuneration). Correlation analysis was conducted to identify relationships between the factors. There were strong correlations ($r > 0.50$) for SRHW, management practices and some HR Outcome items, suggesting strong relationships; however, they do not indicate in which direction the relationship exists (Pallant 2007). Of the eight factors selected through the factor analysis, two (LMX and POS) were used as measures to represent management practices. The remaining six factors represented the HR measures: competence, empowerment, personal isolation, professional isolation, remuneration and work engagement.

E.1.3 Developing the SRHW measure

An *a priori* approach was taken to develop the SRHW measure. Herzberg's motivation-hygiene theory, discussed in chapter five, proposed that the motivation factors are important for job satisfaction; however, the hygiene factors must be addressed to minimise dissatisfaction. The SRWH measure used two questions from the questionnaire to measure hygiene factors, these were *I feel fairly rewarded for the amount of effort I put into my job* and *My employer provides adequately for my personal safety*. As well as using two questions from the questionnaire to measure motivation factors, these were, *My work is satisfying* and *I feel loyal to my employer*.

E.1.4 Identifying Statistically Significant Relationships

The path analysis revealed which HR outcomes had statistically significant relationships. Management practices (LMX and POS) had direct significant relationships with SRHW. There were also statistically significant indirect relationships suggesting that some of the HR outcomes (professional isolation, empowerment and remuneration) moderate the relationship for management practices (LMX and POS) and SRHW. Moreover, while the HR outcomes (personal isolation and competence) had a statistically significant relationship with management practices, they did not have a statistically significant relationship with SRHW, suggesting that management practices in relation to personal support and competence do not influence workforce sustainability. The moderated relationship with remuneration and SRHW had a considerably weaker association than the other HR outcomes suggesting that management practices have a milder impact on the influence of remuneration in improving workforce sustainability.

References

Briggs KK, Lysholm J, Tegner Y, Rodkey WG, Kocher MS, Steadman JR (2009) The reliability, validity, and responsiveness of the Lysholm score and Tegner activity scale for anterior cruciate ligament injuries of the knee. Am J Sports Med 37(5):890–897

Onnis L (2016) A Sustainable Remote Health Workforce: Translating HRM Policy into Practice. Ph.D. Thesis. James Cook University

Pallant J (2007) SPSS survival manual: a step by step guide to data analysis using SPSS. 3rd ed. Allen & Unwin, Australia.

Pelletier LG, Tuson KM, Fortier MS, Vallerand RJ, Briere NM, Blais MR (1995) Toward a new measure of intrinsic motivation, extrinsic motivation, and amotivation in sports: the Sport Motivation Scale. J Sport Exerc Psychol 17(1):35–53

Tabachnick BG, Fidell LS (2007) Using multivariate statistics. 5th ed. Pearson Education, USA

Glossary

Boundaryless career describes a career which is not tied to one employer. Many professional careers are boundaryless, for example, nurses can work for any employer as nursing skills are transferable to any employer. An increase in temporary employment arrangements, reduced job security and worker mobility has lead to the increased popularity of boundaryless careers

Localised management practices are where managers have the skills, capacity and confidence to draw on their own knowledge and experience to find their own situation-specific solutions that are consistent with contemporary practice, within the acceptable boundaries of their profession, and compliant with organisational governance; yet also suitable and appropriate for their remote workplace

Occupational commitment describes an employee's commitment to their profession which usually results from many years of education and personal effort towards the profession and may form part of their self-identity

Organisational citizenship behaviour (OCB) OCB describes the behaviour displayed by employees when they work at a level beyond the expectation for their role for the benefit of their employer

Organisational commitment refers to the strength of an employee's attachment to the organisation and develops slowly, usually after the employee becomes comfortable with the job, the organisation's goals and values, and performance expectations

Protean career is often described as the reinvention of a career. It is where individuals take the skills and experiences gained while working in one career and use them to develop another career. The contemporary nature of work is making this type of career transition more frequently observed, particularly in the service industry

Psychological contract is an unwritten agreement formed by the employee based on their perception of a reciprocal exchange relationship that exists with the employer

© Springer Nature Singapore Pte Ltd. 2019
L. Onnis, *HRM and Remote Health Workforce Sustainability*,
Management for Professionals, https://doi.org/10.1007/978-981-13-2059-0

Psychological contract—breach describes a situation where the employee makes a cognitive evaluation of what they have received compared to what they believe was promised by the employer, and calculates that the employer did not fulfil their obligations

Psychological contract—violation is the emotional response that may follow a perceived psychological contract breach and will most likely result in voluntary turnover

Retention describes the continuance of the employment relationship. While it is not envisaged that the employment relationship is without end, there is a desire that an employee will be retained for a period that benefits the organisation and the employee creating stability until it is mutually deemed a reasonable time period for the employment relationship to end, i.e. more than a few weeks

Voluntary turnover results when an employee decides to resign from their position completely of their own accord. This is differentiated from turnover in general, which includes turnover that is not under the control of the individual employee, e.g. end of short-term contract, compulsory job rotation, redundancy, retrenchment, dismissal, or non-continuance of casual work engagements

Further Reading and Resources for Remote Managers

Books

Giving Voice to Values by Mary Gentile (2010)
This book begins by exploring what 'values' are (and are not), and examines the assumptions beneath workplace behaviours, particularly about why values-based conversations can be difficult. The book then provides guidelines, tools and exercises to help to know 'how' to *voice values*. The power of this book is not in the 'facts', it is in the questions. When faced with an ethical dilemma, the remote manager could use these questions to examine the position, reframe the situation and act in an effective way that is more aligned to their values: http://www.givingvoicetovaluesthebook.com/about/.

Leading on the Edge: Extraordinary Stories and Leadership Insights from the World's Most Extreme Workplace by Rachael Robertson (2014)
In this book Robertson writes about her experience managing a diverse team of people in Antarctica, one of the remotest workplaces in the world. In *Leading from the Edge* Robertson reflects on her experience, drawing examples, emotions and lessons from the reflective journal she kept during her 12 months in Antarctica. The insights that she gains about herself and about managing people in the remote and isolated context are explained at the end of each chapter where she summarises 'what I learned'. This book is not only a convincing read to argue the value of reflective journals; it provides insights that are transferable to other remote and isolated workplaces. Further information and resources for your Manager's Toolkit are available at: http://leadingontheedge.com/.

Mind Reading for Managers by Kim Seeling-Smith (2014)
This book is based on the premise that, most people management issues can be resolved through better communication. A task that is not always simple for managers who often don't know what to talk about with their team and don't know how to have the necessary conversations to motivate, drive performance and improve work engagement. *Mind Reading for Managers* uses social age techniques for social age workplaces.

© Springer Nature Singapore Pte Ltd. 2019
L. Onnis, *HRM and Remote Health Workforce Sustainability*,
Management for Professionals, https://doi.org/10.1007/978-981-13-2059-0

The Game Changer by Jason Fox (2014)
In this book, Fox explains that there isn't a step-by-step guide for change and no magic formulas to motivate people to work. Presented in three sections, the book moves from debunking folklore and conventional management techniques, to designing your 'change game' and concludes with the action of 'changing the game'. The book is easy to read, with relevant sections, signposted for interest and benefits from the amusing images that accompany many of the key points. There are many suggestions and activities in the book that can be implemented in small teams, including teams with limited face-to-face contact. Further information and resources for your Manager's Toolkit are available at: http://www.drjasonfox.com/.

The One Minute Manager by Kenneth Blanchard and Spencer Johnson (1982)
The One Minute Manager is a classic management text. It is a short, easy read with a very clear message for managers about how to work with people, presented as a concise three step method. *The One Minute Manager* provides practical advice about working with others to ensure that they feel good about themselves, achieve the desired results and that a respectful working relationship is established. *The One Minute Manager* provides a new manager an opportunity to build a solid foundation from which further leadership growth is possible. For the established or veteran manager, it provides a refresher into why you chose to be a manager, including a reminder about how valuable human relationships are, especially when developing favourable working relationships.

The Power of Habit by Charles Duhigg (2012)
This book will help remote managers wanting to know more about why some people struggle to change. It provides insight at both a personal and organisational level about how to move towards the change you want to see. *The Power of Habit* may illuminate some of the behaviours observed in remote workplaces. The first section explores 'habits and the individual' helping readers to reflect on their own habits and how habits are shaping their lives. The good news is that once you recognise your habits, you can change—you have a choice! Without a doubt the most valuable parts of the book are the detailed case studies and examples throughout the book. Further information and resources for your Manager's Toolkit are available at: http://charlesduhigg.com/the-power-of-habit/.

The Speed of Trust by Stephen MR Covey (2006)
This book explores *Trust* which Covey describes as being the 'one thing that changes everything'. The book includes detailed explanations of his ideas, examples (often including a story) and tools to aid your understanding and implementation of the ideas generated from this book. There is a good mix of anecdotes and research-based information with a notes and references section at the end of the book for those who would like to investigate further. One of the key messages that Covey expresses is that trust can be increased, much faster than we think; hence the title – *The Speed of Trust*. Further information and resources for your Manager's Toolkit are available at: http://www.speedoftrust.com/

Publications About 'The Study'

The academic publications listed below describe the findings from the study in more detail. They are available from the James Cook University Research Online website (https://researchonline.jcu.edu.au). Please select Onnis, Leigh-ann to go directly to the publications. Weblinks are included where articles are also available directly from the publisher's website.

- Onnis L (2017) Attracting future health workforces in geographically remote regions: perspectives from current remote health professionals. *Asia Pacific Journal of Health Management.* 12(2), 25–33
- Onnis L (2017) HRM Policy choices, management practices and health workforce sustainability: Remote Australian perspectives. Asia Pacific Journal of Human Resources
- Onnis L (2017) What can we learn about improving workforce retention from five words? Proceedings of the 14th National Rural Health Conference, editor Leanne Coleman, Cairns, Queensland, 26–29 March 2017. Canberra: National Rural Health Alliance, 2017 http://www.ruralhealth.org.au/14nrhc/program/concurrent-speakers
- Onnis L (2016) A Sustainable Remote Health Workforce: Translating HRM Policy into Practice. Ph.D. Thesis. James Cook University
- Onnis L (2016) What is a sustainable remote health workforce?: people, practice and place. Rural and Remote Health (Online), 16: 3806 http://www.rrh.org.au/articles/subviewnew.asp?ArticleID=3806
- Onnis L (2016) Attraction and Retention of Health Professionals in Remote Northern Australia: HRM practices in a geographically challenging context. Conference Proceedings. Academy of Management, HR Division International Conference (HRIC), UNSW, Sydney
- Onnis L, Pryce J (2016) Health professionals working in remote Australia: a review of the literature. Asia Pacific Journal of Human Resources 54, 32–56
- Onnis L (2015) An examination of supportive management practices promoting health workforce stability in remote northern Australia. Australasian Psychiatry, 23(6), 679–682
- Onnis L (2014) Managers are the key to workforce stability: an HRM approach towards improving retention of health professionals in remote northern Australia. Conference Proceedings, 28th ANZAM Conference, 3–5 December, UTS Sydney. ISBN:978-0-9875968.

Websites for Networking, Education and Resources

Australian Human Rights Commission

A Quick Guide to Discrimination Law for employers: https://www.humanrights.gov.au/our-work/employers.

Toolkits, Guidelines and other resources: https://www.humanrights.gov.au/employers/toolkits-guidelines-and-other-resources.

Australasian College of Health Service Management
The Australasian College of Health Service Management is the peak professional body for health managers in Australasia. https://achsm.org.au.

Australian Human Resources Institute (AHRI)
AHRI is the peak professional body for HR professionals in Australia. The AHRI website contains information about managing people–articles, podcasts and industry reports. https://www.ahri.com.au/.

Chartered Institute of Personnel and Development (CIPD)
The professional body for HR/people development in the United Kingdom. https://www.cipd.co.uk/.

Chartered Professionals in Human Resources (CPHR)
The national voice on the enhancement and promotion of the HR Profession in Canada. CPHR represents the Canadian HR Profession with HR Associations around the world. https://cphr.ca/.

CRANAplus—Remote Management Program (RMP)
CRANAplus understands the challenges remote managers experience managing staff at a distance, leading change, and being responsible for the delivery of safe, quality care. There are a variety of management and leadership programs; however, the RMP contextualises remoteness and the associated challenges managers experience in remote health services. The RMP is designed to enhance and broaden the manager's existing expertise regarding leadership and management, clinical governance, and project management underpinned by an action learning approach. https://crana.org.au/education/courses/management-course/.

Fairwork Ombudsman (Australia)
Information about workplace rights and obligations. https://www.fairwork.gov.au/.

HRM: Helping Remote Managers
This website acknowledges how time-consuming it is to sift through management books and journal articles so they do it for you! With the remote manager in mind the blogs review contemporary literature and a few classics. They also offer mentoring tailored to the needs of remote managers: www.helpingremotemanagers.com.au.

Human Resource Institute of New Zealand (HRINZ)
Human Resources Institute of New Zealand is the professional body for those involved in Human Resource Management and the development of people. https://www.hrinz.org.nz/.

International Labour Organisation (ILO)
The International Labour Organisation is a United Nations agency that deals with labour problems, including international labour standards, social protection and work opportunities for everyone. http://www.ilo.org/global/lang–en/index.htm.

Singapore Human Resources Institute (SHRI)
Singapore Human Resources Institute is the professional HR body in Singapore. SHRI is committed to high standards of professionalism in human resource management. http://shri.org.sg/.

MIX

Papier | Fördert
gute Waldnutzung

FSC® C083411

Zeitfracht Medien GmbH
Ferdinand-Jühlke-Straße 7
99095 Erfurt, Deutschland
produktsicherheit@kolibri360.de